THE GREAT INDOORS

THE

THE SURPRISING SCIENCE

GREAT

OF HOW BUILDINGS SHAPE OUR

INDOORS

BEHAVIOR, HEALTH, AND HAPPINESS

EMILY ANTHES

SCIENTIFIC AMERICAN
FARRAR, STRAUS AND GIROUX
NEW YORK

Scientific American / Farrar, Straus and Giroux
120 Broadway, New York 10271

Portions of chapters 1 and 8 previously appeared, in different form, on the website of *The New Yorker.* Portions of chapter 4 previously appeared in the September 15, 2016, issue of *Nature.*

Library of Congress Control Number: 2020934822
ISBN: 978-0-374-16663-2

Designed by Richard Oriolo

www.fsgbooks.com • books.scientificamerican.com
www.twitter.com/fsgbooks • www.facebook.com/fsgbooks

10   9   8   7   6   5   4   3   2   1

FOR BLAINE,
MY PARTNER IN ALL THINGS INDOORSY

CONTENTS

THE GREAT INDOORS

INTRODUCTION

I N MITAKA, JAPAN, on a busy street crammed with squat beige buildings, a strange apartment complex jolts the skyline. From the outside, the nine-unit residence resembles a set of child's blocks, with the same kaleidoscopic jumble of shapes and colors—a green cylinder stacked atop a purple cube, a blue cube resting on a yellow cylinder. Inside, the building is an architectural acid trip. Every one of the nine lofts has a circular living room, with a kitchen plopped right in its center. The bedrooms are square, the bathrooms are barrel-shaped, and the studies are complete spheres. Each unit is painted more than a dozen different colors, none of them subtle. (Apartment 302, for instance, has a blue and lime-green kitchen, a lemon-yellow study, and a forest-green bathroom.) Ladders in the living room lead to nowhere. The concrete floors are studded with grapefruit-sized bumps. The building looks less like a home than an oversized carnival fun house. But for all its apparent whimsy, it was designed with a serious purpose: to cheat death.

The Mitaka lofts were created by Shusaku Arakawa and Madeline Gins, married artists who devoted their careers to an idea they called "reversible destiny." Death, they believed, was "old-fashioned," "immoral," and not at all preordained. "That mortality has been the prevailing condition throughout the ages does not mean it will always have to be," Arakawa and Gins wrote in their 2002 manifesto. "Any resistance mounted thus far against mortality, that ineluctable asphyxiator, has been conducted in too piecemeal a fashion . . . The effort to counter mortality must be constant, persistent, and total."

In this effort, they argued, architecture was our most powerful weapon. To resist death, we had to radically reinvent our environments, creating spaces that challenged us both physically and mentally. Living in a place like the Mitaka lofts would keep people off balance, shake them out of their habits and routines, shift their perceptions and perspectives, stimulate their immune systems, and, yes, make them immortal. "We believe that people closely and complexly allied with their architectural surrounds can succeed in outliving their (seemingly inevitable) death sentences!" they wrote.

When I first read about Arakawa and Gins, I assumed it was all an elaborate metaphor, an artistic provocation. But when I visited the Manhattan headquarters of the Reversible Destiny Foundation in the fall of 2018, I learned that they meant it literally. "I think they actually believed that if we achieved this, we could extend our life span," said Miwako Tezuka, the consulting curator of the foundation, which Arakawa and Gins founded in 2010. "They were really, really, really passionate about their belief."

They put this belief into practice, building half a dozen projects on both sides of the Pacific. In Yoro, Japan, they designed a 195,000-square-foot public park so destabilizing that visitors are provided with helmets. In East Hampton, New York, they created the

Bioscleave House, a single-family home even more extreme than the Mitaka lofts, featuring some forty eye-popping paint colors, windows placed seemingly at random, and steep, bumpy floors surrounding a sunken kitchen. "You're going to twist your ankle," warned Stephen Hepworth, director of collections at the Reversible Destiny Foundation. "You may well fall into the kitchen if you're not careful. Don't rush to go to the bathroom."

Although each of their constructions is unique, they're all designed to disorient, with collisions of shapes, colors, and surfaces and sudden shifts in orientation and scale. (In fact, their spaces are so counterintuitive that they come with instructions.) Exiting one of their buildings, Hepworth said, is "like getting off a roller coaster. You're a bit off kilter."

Arakawa and Gins had even bigger dreams, for entire reversible destiny developments, neighborhoods, and towns, or what they described as "cities without graveyards." They wanted to wage a full-scale, all-out architectural war on mortality. But if they discovered the secret to eternal life, they failed to avail themselves of it. Arakawa passed away in 2010 (Gins refused to reveal the cause of death: "This mortality thing is bad news," she told *The New York Times*), and Gins died of cancer four years later.

Their body of work, however, lives on. Those who wish to defy death can rent one of the Mitaka lofts through Airbnb.

THE NOTION THAT ARCHITECTURE can help us live forever is clearly science fiction. But the promise of improving our health and extending our life spans, even just a little, without ever leaving the house? Well, I found that idea irresistible. After all, I am unapologetically indoorsy. It's not that I don't like nature; I think nature is lovely. I've been camping numerous times—and enjoyed it! It's just that I'm anxiety-prone and risk-averse, and the world inside my apartment is warm and cozy and safe. Lots of journalists file dispatches from far-flung places—reporting on wildlife in the Serengeti, floods in the Mekong Delta, and ice cores in Antarctica—but I've always felt most comfortable plying my craft from deep inside my living room.

Though I might be at the extreme end of the spectrum, I am not alone; modern humans are essentially an indoor species. North Americans and Europeans spend roughly 90 percent of their time inside, and the indoor environment dwarfs the outdoor one in some major cities. The island of Manhattan is only twenty-three square miles in size but has three times that much indoor floor space. And unlike the outdoor world, the indoor world is expanding. Over the next forty years, the United Nations estimates, the total amount of indoor square footage will roughly double worldwide. "Those additions are equivalent to building the current floor area of Japan every single year from now until 2060," the organization reported in 2017.

To my delight, more and more scientists have begun to view the indoor environment as worthy of investigation. Researchers in a wide range of fields are now surveying the indoor world, mapping its contours and uncovering its secrets. Microbiologists are charting the bacteria that bloom in our buildings, and chemists are tracking the gases that waft through our homes. Neuroscientists are learning how our brains respond to different building styles, and nutritionists are investigating how cafeteria design affects our food choices. Anthropologists are observing how office design influences the productivity, engagement, and job satisfaction of employees around the globe. Psychologists are probing the connections between windows and mental health, lighting and creativity, and furniture and social interaction.

Their findings suggest that the indoor environment shapes our lives in far-reaching and sometimes surprising ways. To name just a few: Women who give birth in sprawling hospital wards are more likely to undergo cesarean sections than those who labor in more compact ones. Warm, dim lighting makes schoolkids less fidgety and aggressive. Fresh, well-ventilated air boosts office workers' cognitive function.

And the physical location of our homes can have all sorts of ripple effects on our lives. In a 2016 study, a group of Canadian doctors reported that living on the upper floors of a skyscraper can literally be deadly. The doctors examined the medical records of nearly eight thousand adults who'd suffered from cardiac arrests in private homes. The higher up people were when they collapsed, the longer it took paramedics to reach them and the lower their odds of survival; 4.2 percent of patients below the third floor survived their ordeals, compared to less than 1 percent of people above the sixteenth floor. Above the twenty-fifth floor, there were no survivors.

But the first floor's no panacea either. In one study, scientists discovered that elementary school children who lived on the top floors of several Manhattan skyscrapers were better readers than those who lived closer to the ground. What could possibly explain the connection? As it happens, the buildings were situated on a bridge that ran across a major highway, and the constant din of traffic made the units close to the ground significantly noisier than those on higher floors. This noise might have made it more difficult for young children to hear the subtle differences in the small units of sound that make up words, a skill that is critical for reading. Indeed, the children living on the bottom floors scored lower on tests of auditory discrimination, and subsequent research has confirmed that noisy environments can derail language learning.

Even Arakawa and Gins's ideas aren't quite as far-fetched as they seem. We know, for a scientific fact, that the right kinds of challenges can strengthen our bodies and minds. (Start lifting weights and your muscles will swell. Learn to speak a new language and your brain will sprout new connections.) There's no reason that those challenges can't come from our homes. Scientists have known for decades that housing lab animals in stimulating spaces—in the company of other animals and in cages stocked with tunnels, toys, mazes, ladders, and running wheels—is better for their health than confining them to spare, solitary cages. This kind of environmental enrichment can boost animals' immune systems, slow the growth of tumors, make neurons more resistant to injury, and stave off the cognitive decline associated with aging.

There's circumstantial evidence to suggest that engaging environments are good for humans, too. Researchers have found that rates of dementia tend to be lower in cities than in rural areas, for instance. It's hard to say exactly why, but one theory is that urban living is more stimulating and complex, and thus protects the brain. "I think spaces that engage us in multiple ways are probably the ones that we will age healthier in," said Laura Malinin, a cognitive scientist and architect at Colorado State University. In her own research, Malinin has collected some preliminary data suggesting that visually complex rooms may boost the cognitive performance of seniors.*

So Arakawa and Gins weren't completely off track. "I'm not sure about 'reversing' destiny, because I think we shape our own destinies throughout our lives, but I do believe that they're tapping into something," Malinin said. "Which is that the physical environment has a strong—and up till now relatively untapped—potential to help keep us healthy."

I DECIDED TO MOUNT an expedition into the great indoors, to reckon with this world that is entirely of our own making. What is the shape of the indoor universe, and how powerful is its influence? What ecosystems does it contain, and how do we fit into them? How do these interior landscapes shape our thoughts, feelings, and behaviors; our social interactions and relationships; our health, happiness, and well-being?

Finding answers to these questions would require me to venture beyond the walls of my home—at least temporarily. In the chapters that follow, we'll tour an operating room designed to minimize medical errors, an elementary school designed to nudge kids into being more

---

*In some ways, this idea runs counter to the common practice of making environments for seniors less challenging. "A lot of the guidelines for designing for aging are things like neutral colors, simple settings, simple solid surface flooring, beige walls and those types of things," Malinin said. "We create things that are all on one level, that are visually simple, that are easy to navigate. And in so doing, in some sense, we are making environments more impoverished as opposed to enriched."

active, and a prison designed to support inmates' psychological needs. We'll learn how scientists are using brain-wave-measuring headsets, biometric wristbands, environmental sensors, digital mapping, machine learning, and virtual reality to study the built environment and track how people respond to it. And we'll consider how buildings will shape our future, from smart homes that monitor our health to amphibious floating houses that could help us survive climate change. We'll even take a brief, long-distance look at the ice-covered domes we might find ourselves erecting on Mars.

It's time we give the indoor world its due. For too long, we've neglected indoor environments; they've become so familiar to us that we've overlooked their power and complexity. That's finally changing, and the more we discover about our interior landscapes, the more opportunities we have to transform them. Through thoughtful and careful design, we can improve nearly every aspect of our lives. We are products of our environments, but we don't have to be victims of them.

Even small design changes can have dramatic effects. Consider what happened after the Women and Infants Hospital of Rhode Island unveiled a new neonatal intensive care unit (NICU). Traditionally, premature infants born at the hospital had been cared for in big, open wards. These wards were chaotic, crowded, and noisy, filled with beeping machines and constant conversation. On any given day, a dozen infants, many in incubators, were lined up against the walls, and there was little space for parents who wanted to spend time with their babies.

In 2009, however, the hospital opened the new NICU, which did away with the open-bay model; instead, each preemie was assigned to a spacious single-family room equipped with a sleeper sofa where parents could crash for the night. This one change—from communal open wards to private rooms—made a big difference in the babies' development. Infants who spent the first weeks of their lives in the new rooms gained weight more quickly and weighed more at discharge than those who'd been treated in the open bays. They were also less likely to develop sepsis, required fewer medical procedures, and displayed fewer signs of stress and pain.

Architecture isn't the solution to all of our problems. The effects of design interventions are often subtle and complex, and built environment studies can be difficult to conduct and interpret. Moreover, the challenges that the experts in this book are trying to tackle, from preventing chronic disease to making correctional systems more humane, will require much more than infrastructure upgrades. Take that remarkable NICU study. The physical space likely had some direct benefits for the infants; studies suggest, for instance, that noise can derail the development of preemies, increasing their heart rate and blood pressure and decreasing the oxygen saturation of their blood. These physiological responses may partly explain why infants fared better in quiet, private rooms. But the benefits of single rooms can't be attributed to architecture alone. Part of what made the redesign so powerful was that the single-family rooms made it easier for parents to spend time with their infants and be involved in their care.

This is what good design does—it expands what's possible. It nudges us in the right direction, supports cultural and organizational change, and allows us to express our values. Good architecture can help us lead healthier, happier, more productive lives; create more just, humane societies; and increase our odds of survival in a precarious world. It can be the infrastructure on which we build a better future. Even if it doesn't make us immortal.

ONE

THE INDOOR JUNGLE

O N A SUNNY, unseasonably warm afternoon in October, I step into my shower fully clothed. I snap on a pair of blue nitrile gloves, rise onto the tips of my toes, and carefully unscrew my showerhead. Reluctantly, I peer inside. I exhale. It's not nearly as bad as I'd feared. There's no muck, no murk, no layer of overgrown slime. There's not even a single visible speck of dirt. Relieved, I rub the tips of two cotton swabs around the interior and slide the swabs into a thin plastic tube.

Then I sit down at my dining room table to work through a detailed questionnaire about my showerhead: When was it installed? How would I describe its spray pattern? How often do I clean it?

Am I supposed to clean my showerhead? I wonder. *Is that something that people do?*

I circle "Never," seal the survey and the collection tube inside a small white envelope, and drop the package in the mail.

My showerhead swabs are headed to Noah Fierer, a microbiologist at the University of Colorado Boulder who will scour them for signs of hidden life. More specifically, he'll search for microorganisms, also known as microbes, a group of creatures so small that they're typically invisible to the naked eye. It's an umbrella term for all sorts of life-forms, including bacteria—single-celled organisms that are generally shaped like rods, spheres, or spirals—and fungi like yeasts and molds. (Of course, if you've ever stumbled upon a forgotten loaf of bread or an aged hunk of cheese, you'll know that molds do become visible when colonies grow large enough.)

Microbes rule the planet, making themselves at home in nearly all of its habitats. They live on top of Mount Everest and miles below the Earth's surface; in the Namib desert and the Sargasso Sea; in hot springs, storm clouds, deep-sea trenches, lakes of liquid asphalt, the roots of soybeans, the guts of tropical caterpillars, and, of course, ourselves. Our bodies are home to a restless mess of microbes; some are capable of causing disease, but others are crucial partners in maintaining our health. Microbes help digest our food and protect against infection; they keep our metabolisms humming along and our immune systems finely tuned. They even affect our brains, shaping our moods and behaviors. According to the latest estimates, there are roughly as many bacterial cells in our bodies as human ones.

Over the course of his career, Fierer has gone microbe hunting all over the world, mounting expeditions to Panama, New Zealand, and Antarctica. And now he's going to turn his attention to a less exotic location: my showerhead. "It sounds crazy," Fierer admitted when he first told me about the study. "That's the most random environment to sample. But it turns out that there's a lot of bacteria that live in your showerhead." These bacteria mass together in thin, slimy layers known as biofilms. (Biofilms aren't limited to the showerhead—they can attach to all sorts of surfaces, including river rocks, medical implants, and teeth. Dental plaque, for example, is a biofilm.)

And what happens in the showerhead doesn't stay in the showerhead; when a jet of hot water comes blasting through, some of the microbes find their way into the spray. "And then you breathe that in directly," Fierer said. "I think it's a really important mechanism by which we're exposed to bacteria." But a few years ago, it occurred to Fierer that scientists didn't know precisely which species we were inhaling whenever we stepped into the shower. So he decided to find out. Working with Rob Dunn, an ecologist at North Carolina State University, he set out to scour hundreds of showerheads across the United States. Together, they'd inventory the microbial species hiding in each showerhead, analyze how they varied from home to home, and start to unpack how these critters might affect us.

The study is an outgrowth, so to speak, of the burgeoning field of indoor ecology. Fierer and Dunn are part of an intrepid tribe of indoor explorers who have begun to survey the invisible menagerie of organisms that inhabit our homes. "We've just opened up this giant black box of what lives with us," Dunn told me. There's far more to our homes than meets the eye; even the most sparkling house contains vibrant, unseen ecosystems. The emerging research suggests that the lives of these organisms are inextricably intertwined with our own— and that being more mindful of these creatures could help us create a healthier house.

The prospect is both alluring and unnerving. The more I read about the world of indoor microbes, the more I found myself obsessing over my own invisible roommates. I contemplated fungi as I cooked, bacteria as I bathed. I began to feel like a stranger in my own home, humbled by how little I knew about what was happening under my roof. I decided that the time had come to get to know my microbes, so

I got to work swabbing my bathroom and then set off for Colorado to meet the man behind the shower curtain.

I ARRIVED AT the University of Colorado Boulder in early January, during the first week of classes after the winter break. Students streamed across the quad as Fierer—rugged and ruddy-faced, a bike helmet tucked under his arm—walked me to his office in the environmental science building. "So this is where the magic happens," he said, gesturing around his light-drenched second-floor laboratory. The four large freezers that sit against the back wall are stuffed with samples: soil from Colorado, moss from Alaska, caterpillars from Costa Rica, all brimming with microbes.

Fierer found his calling by process of elimination. After graduating from college with degrees in biology and art history, he bounced between research jobs. He worked with salamanders and birds, and spent two years trapping wild gerbils in the Israeli desert. He hated it: "They were disgusting and trying to bite me, and I was like, 'I don't want to work with animals.'" So he tried conducting tree surveys on the Oregon coast. "I like plants, but I didn't find them that compelling," he acknowledged. And just like that, the aspiring ecologist had eliminated both flora and fauna as subjects of further study.

When Fierer started graduate school in the late 1990s, he decided to think smaller. He started studying soil—and the microbes that dwell there, breaking down organic matter and recycling its nutrients. His timing was good; advances in DNA sequencing technology were about to break the field of microbiology wide open.

Though bacteria don't bite, they present their own research challenges. Many don't grow well—or at all—in the laboratory. The emergence of gene sequencing provided a powerful new way to identify them, allowing scientists to collect a sample of soil or water and sequence all the DNA it contained. Then they could match these sequences to known bacterial and fungal genomes, generating a snapshot of the microbial species that were present. As gene sequencing became easier, cheaper, and faster, many microbiologists employed the technique to inventory the organisms living in all sorts of outdoor environments, from Arctic ice floes to thickets of Amazonian jungle. But a small group of scientists began to wonder what they'd find if they looked a bit closer to home—literally. "We spend a lot of time indoors," Fierer told me. "And a lot of the organisms that we encounter on a daily basis are those that are inside our homes."

In 2010, Fierer made his first foray into the indoor microbial world, cataloging the bacteria present in twelve campus restrooms.* The following year, he studied the microbes in residential kitchens and partnered with Rob Dunn to launch the Wild Life of Our Homes project. They began with a small pilot study in North Carolina, recruiting forty families to run cotton swabs across seven surfaces inside their homes: a countertop, a cutting board, a refrigerator shelf, a pillowcase, a toilet seat, a TV screen, and the trim around an interior doorway.

The homes were crawling with microbial squatters—more than two thousand types, on average. Different locations within the homes formed distinct habitats: kitchens harbored bacteria associated with food, while doorways were covered in species that typically live in leaves and soil. From a microbiological perspective, toilet seats and pillowcases looked strikingly similar; both were dominated by bacteria that typically live on our skin and in our mouths.

Beyond these commonalities, there was a lot of variation among the homes, each of which had its own microbial profile, sheltering a slightly different collection of organisms. But the researchers couldn't explain why. So Fierer and Dunn launched a second study, asking more than one thousand families living across the United States to swab the dust that had collected on the trim around their interior doorways.

---

*Among the findings: "Interestingly, some of the toilet flush handles harbored bacterial communities similar to those found on the floor," Fierer and his colleagues wrote, "suggesting that some users of these toilets may operate the handle with a foot (a practice well known to germaphobes and those who have had the misfortune of using restrooms that are less than sanitary)."

"We focused on that because nobody ever cleans it," Fierer told me. "Or we don't clean it very often—maybe you're an exception." (I am not.) Because the dust collects over months or years, the duo hoped it would give them the broadest possible look at indoor life, an inventory of the organisms that had floated, crawled, and skittered through the homes over the previous months and years. As Dunn put it: "Each bit of dust is a microhistory of your life."

Back in the lab, the team analyzed the DNA fragments present in each dust sample, listing every organism that made an appearance. The numbers were staggering. In total, the indoor dust contained DNA from more than 116,000 species of bacteria and 63,000 species of fungi. "The shocker was the diversity of fungi," Dunn told me. There are fewer than 25,000 species of named fungi in all of North America, which means that our houses could be teeming with organisms that are essentially unknown to science. In fact, when the researchers compared the indoor dust to samples that the volunteers had taken from the trim around an exterior door, they found that there was more microbial diversity inside the homes than outside of them.

Some of the species that Fierer and Dunn identified originate outside, hitching rides into our homes on our clothes or drifting in through open windows. (And they may not all be alive by the time they turn up inside; DNA sequencing can identify the organisms that are present in a sample, but it can't distinguish between living creatures and dead ones.) Other kinds of bacteria actually grow *in* our homes—in our walls and our pipes, our air conditioning units and our dishwashers. Some sprout on our houseplants or our food.

And a lot of indoor microbes, it turns out, are living on us. "We're constantly shedding bacteria from every orifice and body part," Fierer said. "It's nothing to be grossed out about. It's just the way it is." Our individual microbiomes—the collection of microorganisms that live in and on our bodies—are unique, and we each leave our own microbial signatures on the places we inhabit. In one innovative study, researchers tracked three families as they moved into new homes; each family's distinct blend of microbes colonized its new residence within *hours*. The scientists—led by Jack Gilbert, a microbial ecologist then at the University of Chicago—could even detect the individual microbial contributions of each family member. "People who spent more time in the kitchen, their microbiome dominated that space," Gilbert explained. "People who spent more time in the bedroom, their microbiome dominated there. You could start to forensically identify their movement."

Indeed, the bacteria that turn up inside a home depend enormously on who lives there. Fierer and Dunn found that *Lactobacillus* bacteria, which are a major component of the vaginal microbiome, were most abundant in homes in which women outnumbered men. When men were in the majority, different bacteria thrived: *Roseburia*, which normally live in the gut, and *Corynebacterium* and *Dermabacter*, which both populate the skin. *Corynebacterium* is known to occupy the armpit and contribute to body odor. "Maybe it means that men's houses smell more like armpits," Dunn ventured. "Microbially, that's a fair assessment." The findings may be due to sex differences in skin biology; men tend to have more *Corynebacterium* on their skin— and to shed more skin microbes into the environment—than women do. (The researchers also acknowledge the possibility that a bachelor pad's bacterial profile could be the result of "hygiene practices.") In a subsequent study, Fierer and his colleagues showed that they could accurately predict the sex of the students living in a college dorm room simply by analyzing the bacteria in its dust.

Meanwhile, dogs introduce their own drool and fecal microbes into a home and track soil dwellers in from outside. (Dog owners never seem too bothered when Dunn tells them that Fido is smuggling an entire microbial zoo into their homes. "It's a pretty fine conversation most of the time," he told me. On the other hand, he noted, "If I say that every time your neighbor comes over, that he brings over a mix of beneficial microbes and pathogens, it just makes people scrub.") Cats change a home's microbial makeup more modestly, perhaps because they are smaller and venture outside less often. Using the dust DNA alone, Fierer and Dunn were able to predict whether a home contained a dog or a cat with roughly 80 to 90 percent accuracy.

While the bacteria in our homes mostly comes from us (and our pets), the fungi are another story. Fungi are much less abundant in our own microbiomes, and our houses are dominated by fungal species that originate outdoors. A home's fungal signature, Fierer and Dunn found, was largely determined by where it was located. Houses in eastern states had different fungal communities than those in western ones. Ditto homes in humid climates compared with those in dry ones. The geographic correlation was so strong that Fierer and Dunn could use fungal DNA to determine, to within about 150 miles, where a house dust sample originated.

Fierer and Dunn did identify more than seven hundred kinds of fungi that were more common indoors than out, including a variety of household molds, yeasts, edible mushrooms, and fungi that live on human skin. Homes with basements had different fungi than those without them. And because some species of fungi feed on wood and other building materials, what our homes are made of affects the fungi that live there. "It's kind of a 'three pigs' thing," Dunn told me. "A stone house feeds different fungi from a wood house from a mud house. Because unlike the bacteria, they're eating the house."

BUILDING BY BUILDING, species by species, study by painstaking study, we're beginning to take the measure of indoor microbes and fill in the maps of their expansive empires. We've found them in every possible indoor habitat, including classrooms and offices, gyms and public restrooms, hospitals and airplanes. There are black yeasts in the dishwasher, heat-resistant bacteria on the International Space Station, and foot-associated microbes all over the New York City subway. "That is probably due to the fact that every time you take a step, your heel comes up and then presses down, creating a small bellows of bottom-of-your-foot air squirting out into the surroundings," said Norman Pace, a pioneering microbiologist who led the subway study. "And now just imagine millions of people running around down there. *Puff puff puff puff puff*—every time they take a step, they put out a little puff of foot microbiology."

The research makes clear that there is no "typical" indoor microbiome, and even the most basic design decisions matter. "How you arrange collections of spaces, which spaces are next to other spaces, which spaces are remote from other spaces—those decisions, which are the bread and butter of architects, make a difference in terms of shaping the microbiome," said Jeff Kline, a senior research associate at the Biology and the Built Environment (BioBE) Center at the University of Oregon.

When a group of BioBE researchers collected dust from a four-story building on campus, they learned that spaces that were centrally located and highly occupied, such as hallways and classrooms, had different bacterial populations than farther-flung, more lightly occupied areas, such as faculty offices and mechanical rooms. And the more connected two rooms were—that is, the fewer doors a visitor had to walk through to get from one room to another—the more similar their microbial profiles. The microbes also varied according to room size and building floor.

BioBE scientists have also determined that sunlight shining in through windows can inhibit the growth of some of the bacteria in indoor dust and that rooms that have operable windows, or are naturally ventilated, tend to be home to more plant, soil, and water microbes than those that are mechanically ventilated, which are dominated by microbes from the human body.

Indeed, as our architectural choices have evolved—from homes that were relatively open to the outdoors to those that are walled off from it—so has the indoor microbiome. "Houses are really a gradient of how many environmental microbes we invite in," Dunn said. In a 2016 study, an international team of scientists sampled homes in four communities in the Amazon River basin: a remote jungle village, a rural village, the Peruvian town of Iquitos, and the über-urban Brazilian city of Manaus.

Residents of the jungle village lived in large thatched huts made of wood and reeds. The huts were completely open, with dirt floors and no exterior or interior walls. In the rural village, the homes had external walls but few interior partitions. They had no indoor toilets and were generally built of wood, thatch, brick, and tin. Both kinds of village homes were chock-full of environmental bacteria, including species associated with soil, water, and insects. The town and city homes, on the other hand, had outer walls made of brick, tin, or cement; indoor bathrooms; and interior walls acting as barriers between rooms. And they were covered in bacteria that typically live in and on the human body.*

Even in tightly sealed buildings, however, the microbiome isn't static—it can change as residents come and go or environmental conditions shift. Dampness can encourage the growth of both bacteria and fungi, while a rigorous cleaning regimen can reduce microbial abundance and diversity, at least temporarily. "It's a really dynamic system," Fierer said.

OUR BUILDINGS ARE RICH biological wonderlands, and there's more to them than microbes. Every new discovery that indoor ecologists make highlights how much biodiversity is living right under our noses—and how much is left to discover. For instance, we know so little about the insects that inhabit our homes that even a basic survey can yield major surprises, as Dunn himself learned when he enlisted more than two thousand volunteers from across the United States to scour their homes for camel crickets. About 150 species of these crickets are native to North America; most live in forests, but a few reside in basements and cellars, and Dunn wanted to determine how widespread they were in American homes.

He wasn't terribly surprised to discover that camel crickets were

----

*Animal dwellings are even more extreme, Dunn found when he collaborated with biological anthropologists to swab chimpanzee nests in western Tanzania. "Chimp nests are all environmental microbes," he told me. "You can't tell a chimp has ever been there."

common, especially in the eastern United States, where they were present in 28 percent of residences. What did shock him was how few of the crickets were native. Among all the homes with verifiable camel cricket occupations, just 12 percent had North American species. The rest housed invaders, Asian crickets that had somehow made their way across the Pacific. These Asian camel crickets weren't known to live in our homes, but at some point they'd moved in, en masse and completely undetected.

Camel crickets aren't even the half of it. In 2012, Dunn's team ventured into fifty houses in North Carolina, intent on documenting every single arthropod—a group of invertebrates that includes insects, spiders, and centipedes—they could find. They ended up collecting more than ten thousand specimens, representing more than five hundred species, with almost one hundred species per house, on average. They unearthed spiders, silverfish, springtails, bristletails, earwigs, crickets, cockroaches, termites, centipedes, millipedes, wasps, ants, bees, beetles, moths, fleas, mites, lice—and more.

They discovered carpet beetles, cobweb spiders, gall midges, and ants in every single house. They found book lice in all but one. They turned up predators, parasites, and scavengers; insects that were feeding on animal carcasses, dog kibble, and fingernail clippings. Some of the beetles and flies that they found have previously turned up in archaeological sites, suggesting that they've been sharing our homes for millennia. "There are a bunch of species that have been in houses for thousands of years that we've just not paid very much attention to," Dunn told me. "That's the piece that most excites me."

Dunn has since gone on to catalog the arthropods living in homes in Peru, Sweden, Japan, and elsewhere, hoping to understand the factors that shape their presence and distribution. He and Fierer made some preliminary discoveries simply by sequencing the arthropod DNA in the house dust of American homes. Rural homes had more diverse collections of arthropods than suburban or urban ones; homes with dogs, cats, or basements also had particularly rich assortments of arthropod life. "Basements are like hotbeds of arthropod diversity," Fierer noted.

Some organisms come and go from our homes, lured in by food scraps or electric light. "Your home is essentially acting like a gigantic light trap," Fierer explained. Others spend their entire lives there, and our buildings are home to some singular forms of life. Bedbugs and German cockroaches are essentially confined to human dwellings, and the black yeasts that grow in bathrooms and washing machines appear to be genetically distinct from those that grow on soil and decaying leaves. "The new habitats that we create in homes seem to offer up new fungal niches," Dunn explained. And Fierer recently discovered a novel virus in the HVAC filters of campus dorm rooms.

So our homes aren't just ecosystems, they're unique ones, hosting species that are adapted to indoor environments and pushing evolution in new directions. Indoor microbes, insects, and rodents have all evolved the ability to survive our chemical assaults, developing resistance to antibacterials, insecticides, and poisons. (German cockroaches are known to have evolved a distaste for glucose, which is commonly used as bait in roach traps.) Some indoor insects, which have fewer opportunities to feed than their outdoor counterparts, seem to have evolved the ability to survive when food is scarce. Dunn and other ecologists have suggested that as the planet becomes more developed and more urban, more species will evolve the traits they need to thrive indoors. (Over a long enough timescale, indoor living could drive our evolution, too. Perhaps my indoorsy self represents the future of humanity.)

We've created and shaped these ecosystems, but we are also part of them, and they influence us back, affecting our health and well-being. The cockroaches and dust mites that live in our homes can trigger allergies. The flies, ticks, and mosquitoes can carry disease. (Fierer and Dunn found *Rickettsia* bacteria—which live inside ticks, fleas, and lice and can cause ailments ranging from typhus to Rocky Mountain spotted fever—in a number of indoor dust samples.) These pests can have psychological effects as well; cockroach infestations may raise the risk of depression, a 2018 study of public housing residents suggests.

But other indoor arthropods could protect us from illness. In Thailand, house spiders feed on mosquitoes that carry dengue fever; in Kenya, they devour malaria-carrying mosquitoes. Some African and Latin American societies have long recognized the benefits of indoor arachnids, deliberately bringing spiders into their homes as a form of natural pest control.

In addition to all this fauna, our homes are full of flora; even if you don't own a single ficus or fern, pollen and plant matter drift in from outdoors. Homes in the Pacific Northwest contain lots of moss and cypress DNA, Fierer and Dunn found, while those in the Southeast are rich in DNA from warm-weather grasses. Approximately 8 percent of the plant species they detected indoors are known to produce allergens. "We always think of pollen as just being outside, but there's a lot of pollen inside the home," Fierer said. "Every time you walk across your carpet, you're kicking up stuff that came from a tree four months ago."*

More indoor dangers lurk if you peer beyond the bounds of biology. Lead remains a major public health concern, and flame retardants—which have been linked to cancer, neurodevelopmental delays, and hormonal problems—soak many of our household goods, from our sofas to our TVs. Many of the basic activities we perform regularly in our homes, like cooking and cleaning, produce gases and airborne particles that are dangerous when inhaled. In one recent experiment, a team of Colorado researchers concluded that preparing a full Thanksgiving dinner can send a home's air quality index soaring above two hundred, into the "very unhealthy" zone. (Although houseplants could

---

*Some of the other plant species that popped up in the indoor dust samples were more curious. "We pick up coffee DNA," Fierer said. "We pick up olives." They identified genes from rice and tea and bananas. These plants probably aren't growing in or around any homes in the continental United States—but they are eaten in them. And whenever you brew a pot of coffee or drop a banana into the blender, some of those food particles can end up on your countertops, floors, and, yes, on the trim around your doorways. It's a record of your meals, written in your dust. "In theory we could figure out what you're eating from looking at plant DNA in the dust from your door trim in your living room," Fierer told me. "I'd love to do a study where you sample dust from different restaurants and figure out if you can reconstruct the menu."

theoretically help purify indoor air, in practice, it's difficult to stock your home with enough of them to make a meaningful difference.)

So if a single roast turkey can irritate my lungs, a smidge of pollen-laden house dust can make me sneeze, and a few insects can make me itch, what could billions of indoor bacteria possibly be doing to me?

SOME OF THE microbes that inhabit our homes are known to cause disease. Black mold, which grows in and on our walls, can trigger allergies and respiratory problems. *Aspergillus fumigatus*, a fungus that can cause lung infections in people with weakened immune systems, lives in our pillows. *Legionella pneumophila*, a bacterium that causes Legionnaires' disease, loves indoor plumbing. It nestles inside hot water tanks, cooling towers, and faucets, and spreads through airborne, or aerosolized, droplets of water. *Streptococcus* bacteria—which can cause strep throat, sinus and ear infections, pinkeye, meningitis, and pneumonia—are more abundant inside our homes than outside them, Fierer and Dunn found. Though the mere presence of these microbes isn't necessarily dangerous, and not all strains cause illness, buildings can provide an infrastructure that helps diseases spread. Airborne influenza can waft through an office building's ventilation system; a spray of *Streptococcus* can turn a doorknob into a booby trap.

But many indoor microbes are completely innocuous, and some may even have lifelong health benefits. In recent decades, the rates of asthma, allergies, and autoimmune diseases have skyrocketed in industrialized nations. Some scientists have theorized that the increasing prevalence of these diseases may be the fault of our modern lifestyles, which keep us at a distance from the robust microbial menageries that surrounded our ancestors for most of human evolution. As a result, our immune systems never get properly trained.*

---

*This theory initially became known as the "hygiene hypothesis," but that terminology is misleading, suggesting that basic sanitation and hygiene practices are to blame. (Please: Don't stop washing your hands.) It's not that our environments are "too clean," exactly—it's that a wide

Evidence has been accumulating to support this theory. Studies show that children who live with dogs, which increase the richness and diversity of bacteria in a home, are less sensitive to allergens and less likely to develop asthma. (A dog might be the immune system's best friend.) Children who grow up on farms, and are exposed to livestock and their microbes, appear to be similarly protected from allergies and asthma.

Some of the most compelling evidence comes from research on two American farming communities: the Amish and the Hutterites. Although the groups have much in common—including large families and Central European ancestry—just 5 percent of Amish kids have asthma, compared to 21 percent of Hutterite children. The communities also have distinct farming customs. The Amish, who generally eschew electricity, live on single-family farms and employ traditional agricultural methods, using horses to plow their fields. It's not uncommon for Amish children to play in the family barns, which are typically located near their homes. The Hutterites, on the other hand, live together on big, industrial farms, complete with high-tech tools and equipment, and their children have less contact with livestock.

These differences may affect the children's microbial exposures and the development of their immune systems. In 2016, scientists reported that house dust collected from Amish households had higher levels of endotoxins—molecules contained in the cellular membranes of some bacteria—than dust from Hutterite homes. What's more, when they drew blood from kids in both communities, they found that compared to Hutterite children, Amish children had more neutrophils, white blood cells that help the body fight infection, and fewer eosinophils, which play a critical role in allergic reactions.

The researchers also whipped up some house-dust cocktails, mixing dust samples from Amish and Hutterite homes with water, and

variety of factors, including urbanization, decreasing family size, and the widespread use of antibiotics, mean that we're no longer exposed to the same rich assortment of microbes in early life. As a result, many scientists now prefer to describe this theory as the "old friends hypothesis."

then shooting the slurries into the nasal passages of young mice. Then they exposed the mice to allergens. The mice that had received the Hutterite dust responded as expected; their airways trembled and twitched. But the mice that had received the Amish dust continued to breathe relatively freely, seemingly protected from this allergic response.

Although there's still a lot to learn, the science suggests that a healthy home is one that's full of uninvited guests. "We are exposed to microbes every day, and a lot of these are harmless or potentially beneficial," Fierer told me. "We don't want a sterile house." Which is good, because it turns out that I don't have one.

A YEAR AFTER I plumbed the depths of my showerhead, an e-mail landed in my inbox: "Your Showerhead Data Are Here!" I nervously clicked on the message. Despite all that I'd learned about the ubiquity of microbial life, I'd spent the previous months fretting; my showerhead had looked so absolutely spotless that I thought Fierer might not find any microorganisms at all. What if my superlatively clean showerhead threw off the whole study?

I should have known better. My showerhead hosted a stunningly diverse array of organisms. It was brimming with *Bradyrhizobium*, a group of bacteria commonly found in soil and tap water, and *Sphingomonas*, rod-shaped bacteria that can break down some common pollutants. It also contained traces of a few more mysterious tenants, including an organism called RB41, which has been found in dog noses and paleolithic cave paintings, and another known as MLE1, which is related to blue-green algae but gets its energy from carbohydrates, rather than sunlight. What was MLE1 doing in my bathroom?

"I have no idea," Fierer admitted. "It was just discovered a few years ago, and nobody's been able to grow it in a lab, so we don't really know what it's capable of doing. But in many of these showerhead samples, it's quite abundant, which I think is interesting, because it's not an exotic environment. We're talking about the showerhead in people's homes, yet we can find these major groups of bacteria that haven't really ever been studied, or not studied well."

In my showerhead, one class of microbes reigned supreme: myco-bacteria, a family of organisms that encompasses almost two hundred species. Mycobacteria are hardy suckers, unbothered by hot water and chlorine, and when they're inhaled, they can cause some really nasty diseases, including tuberculosis, leprosy, and what are known as non-tuberculous mycobacterial lung infections, which are becoming more common in the United States and elsewhere. In fact, Fierer and Dunn discovered that showerheads harboring potentially dangerous strains of mycobacteria clustered in the same regions—Hawaii, Southern California, Florida, and the mid-Atlantic—that are known to be hot spots for mycobacterial-related respiratory infections. "There's been this debate about where people are getting this disease from," Fierer said, "and this suggests to us that they may be getting it from showerheads."

This did not strike me as good news, given that mycobacteria made up 67 percent of the bacteria in my own showerhead. "Nothing to freak out about," Fierer said. "Many of those could be totally non-pathogenic." It's not the most reassuring thing that anyone has ever said to me, but Fierer pointed out that mycobacteria are a complex and interesting group of organisms. "There's been some nice work on that showing that some of these bacteria, when you inhale them, they can actually boost your immune system," he said. Fierer and his colleagues hope to start untangling which are which by exposing mice to some of the specific mycobacterial strains they've found in Americans' bathrooms.

Until we can separate the good from the bad, it's hard to know what to make of my results. Fierer and Dunn did discover that metal showerheads like mine tend to have more mycobacteria than plastic ones, but it's not clear that I'd derive any real benefit from making the switch. Fierer urged me not to panic and to put my results in perspective. Being able to fuss over the mysterious mycobacteria in my showerhead is a luxury. In many parts of the world, water carries much more dangerous organisms, like the bacteria that cause cholera, and access to clean water is not guaranteed even in the United States. (Just ask the people of Flint, Michigan, who will be dealing with the fallout from lead-contaminated water for years to come.)* I'm lucky that my water is so clean. "I don't want to be peddling paranoia," Fierer said. "The last thing I want people doing is tossing out their showerheads every three months because they're worried about bacteria."

SO I'LL KEEP my bathroom hardware for now, but are there other things I can do to engineer a healthier home microbiome? In theory, we should be able to cultivate our indoor microbial gardens, weeding out the dangerous species and helping the healthful ones flourish. "In the future, it may be possible to design buildings that sustain healthy microbiomes," the National Academies of Sciences, Engineering, and Medicine wrote in a 2017 report.

The absolute best way to keep a home healthy is to keep it dry. A lot of the fungi hanging out in our homes are essentially dormant as long as they don't get wet. But if there's a flood or a leak or even just some excess humidity, they spring to life and begin to spread. In one alarming study, Danish researchers discovered that brand-new panels of drywall, purchased at four different shops in Copenhagen, came "pre-contaminated" with several kinds of fungi, including black mold. When they soaked the drywall in sterile water, the fungi started to grow.

Beyond keeping moisture at bay, boosting the rate of air ventilation can help eliminate potential pathogens and contaminants. Removing carpeting—which harbors dust, dander, and debris—can reduce the concentration and persistence of allergens indoors. "From a microbial perspective, carpeting is disgusting," Fierer said.

We'd also be well advised to lay off household products that are ex-

---

*It is not a coincidence that Flint is a low-income community. The burden of bad indoor environments falls mainly on the poor.

pressly designed to kill microbes. "A lot of what we put in the built environment is antimicrobials," said Erica Hartmann, a microbiologist at Northwestern University. "We put them in building materials, we impregnate cutting boards, any kinds of plastics, tiles, paint, all kinds of stuff. We're using these antimicrobials everywhere."

Bacteria adapt to these chemicals at lightning speed, and using them in our homes could help drive the emergence of new superbugs. Hartmann has discovered a correlation between two commonly used antimicrobial compounds and several genes that are known to make bacteria resistant to antibiotics; the more concentrated the chemicals were in a sample of indoor dust, the more abundant the resistance genes were. "It doesn't mean that exposure to antimicrobials is what's making these bacteria more antibiotic resistant, but it is suspicious," Hartmann told me. "It's a sign that maybe we should take another look at what we're doing with antimicrobials in the built environment."

Moreover, coating the inside of our homes with antimicrobial compounds can wipe out the good microbes along with the bad. In any quest to eliminate pathogens from our buildings, we want to spare the species that could be good for our health. Which leads to another enormous challenge: we're still not quite sure which species those are. "The field of medicine is really good at identifying germs," Fierer said. "It's not so good at identifying beneficial microbes."

That hasn't stopped companies from hawking home probiotics, household cleaners, air purifiers, and sprays that will coat your living quarters with a fine mist of what are supposedly beneficial strains of bacteria. The manufacturer of one probiotic spray claims that it "restores the balance of healthy bacteria" and "creates a healthy immune system defense for your home." But few of these products have been rigorously tested, and clinical trials of other kinds of probiotics, like those that are taken orally, have yielded mostly disappointing results. Even if scientists manage to identify effective probiotics, there are probably better ways to deliver them than spraying them around the house. "It just seems like the least efficient possible delivery," said

Brent Stephens, who leads the Built Environment Research Group at the Illinois Institute of Technology. "When we take a vitamin, we don't just squirt it into the air and walk around it."

Moreover, it's extremely unlikely that we're going to discover one magic microbe, a single ideal organism that helps us ward off a cold or stops hay fever in its tracks. There's not even an ideal microbiome— swab one hundred healthy people, and you'll find one hundred different microbial mixtures. A species that promotes health in one person might cause sickness in another; a microbe that benefits a developing child might be dangerous for a senior citizen. So designing buildings that promote healthy microbiomes, when we don't know what we're aiming for? "It's like stomping on the accelerator without knowing what direction you're pointed," said Rob Knight, a microbiome researcher at the University of California, San Diego.

For the time being, the best way to foster a healthy home microbiome requires no fancy products or technology. Keep things dry. Forgo cleansers, textiles, and materials that contain added antimicrobials. Open a window. Get a dog. (Or, if you can swing it, a cow.)

Most of all, we should get comfortable with the incontrovertible fact that we're outnumbered in our own homes. Our buildings are alive, and even their tiniest inhabitants can have profound effects on our well-being. That can seem unsettling, but it also represents an opportunity— to craft indoor spaces that actually improve our health. There's no better place to learn that lesson than in hospitals, where design can be a matter of life and death.

A HOSPITAL ROOM
OF ONE'S OWN

IN FEBRUARY 2013, the Center for Care and Discovery, a ten-story hospital in Chicago, officially opened its doors. As the first patients began to stream in, they brought their microbes with them. They shed bacteria in the lobby, sprinkled viruses around the hallways, deposited fungi in their beds. And they shared these microorganisms with their fellow patients, passing them along to subsequent occupants of their rooms. "When a patient moved into a new room, they were actually colonized briefly from some of the bacteria in the room—the previous occupant's bacteria," said Jack Gilbert, the microbial ecologist who led a yearlong study of microbes in the new hospital. "And that was true even if the room had been cleaned."

After one day, however, the flow of microbes reversed, streaming from the patient's body to the surfaces in the room. Within twenty-four hours, the microbes on the bedrail, the faucet, and other surfaces closely resembled those that the patient had brought in with him. "There's a very rapid turnover," Gilbert said. After the patient was discharged, the cycle would repeat itself, with the room's new resident at first acquiring the previous patient's microbes, and then sprinkling his own microorganisms around the space, in an endless game of microbial telephone.

This microbe swapping happens in all kinds of buildings, but in hospitals, where many people harbor pathogens, it can be especially hazardous. SARS, the deadly respiratory virus that emerged in China in 2002, spread in hospitals and emergency rooms, where patients infected one another and the clinicians who were caring for them. Pathogens can persist even after the patients who deposit them are discharged; when one hospitalized patient suffers from a *Clostridium difficile* infection, which can cause severe diarrhea and even death, subsequent occupants of the room are at increased risk for developing the same affliction.

Many inpatients have weakened immune systems or open wounds, leaving them vulnerable to infection. The spread of antibiotic-resistant strains of bacteria and fungi is making these health-care-acquired infections, which affect 7 to 10 percent of patients worldwide, more dangerous and difficult to treat.

These challenges have prompted health-care architects to start designing with microbes in mind. When administrators at the Skåne University Hospital in Malmö, Sweden, decided to rebuild their department of infectious disease in 2005, they tried to create a building that could operate safely in what they called the "postantibiotic era"—an age in which effective antibiotics are disappearing and epidemics can travel around the world at lightning speed.

To keep the sharing of space to an absolute minimum, the planning team decided that every patient would have a private room, which is known to reduce the spread of infectious disease. (The effect can be dramatic. When Montreal General Hospital switched from shared to single ICU rooms, the rates at which patients acquired potential pathogens, including several drug-resistant strains of bacteria, fell by more than 50 percent, and the average length of stay declined by 10 percent.)

But the design team went further than that—they didn't even want patients passing one another in the hallways. So they created a circular building with balconies that wrapped all the way around the patient wards on the upper floors. Each patient room has two sets of doors, one that opens into a corridor inside the hospital, which is used primarily by staff, and another that leads directly to the outdoor walkways. Patients enter their individual rooms through these outer doors. "You can take patients from the outside directly to their room, so they don't sit in waiting areas coughing and having fevers," said Torsten Holmdahl, who was the head of the infectious disease department and involved in the planning process. (The outpatient clinic and emergency department, on the first floor, also have entrances that lead directly from the outside of the hospital into private examination rooms.)

Both the interior and exterior entrances open onto small anterooms, where staff and visitors can wash and disinfect their hands and don masks and gowns, if necessary. (Though the evidence is mixed, some studies suggest that providing conveniently located sinks and hand disinfectant can improve staff hand hygiene, reducing the odds that clinicians transfer bacteria from one patient to another.)

The anterooms, which have airtight doors, are also pressurized, which keeps contaminated air from flowing into them. "It protects the patient from the outside and it protects the outside from the patient," Holmdahl said. The deliberately oversized patient rooms can be transformed into double rooms in the event of an outbreak or epidemic, or converted into high-risk isolation rooms by bumping up the ventilation rate and locking the anteroom doors.

The building, which opened in 2010, has been working well overall, and disease seems to spread less readily than it did in the old facility, Holmdahl told me. Though scientists haven't formally analyzed patient outcomes, the redesign is a harbinger of a future in which architects take microbial life seriously. And it's fitting that it's happening in hospitals, the birthplace of a discipline known as "evidence-based design."

Over the last few decades, researchers have assembled an overwhelming body of evidence that hospital design affects patient outcomes. In hospitals, they found, architecture can literally save lives; the right design decision can decrease stress and alleviate pain, improve sleep and elevate mood, reduce medical errors and prevent patient falls, curb infections and speed recovery. Thousands of studies have now made it abundantly clear: good design is powerful medicine.

THE MODERN HOSPITAL is the most recent manifestation of an idea that dates back centuries; throughout history, many societies have created their own unique structures for ministering to the sick. The ancient Greeks built temples where ailing citizens could seek guidance from Asclepius, the god of healing, while the Romans created *valetudinaria*, military hospitals for ill and injured soldiers. In medieval Europe, health care was often intertwined with religion; monasteries operated infirmaries, and clergy members ran freestanding Christian hospitals.

Secular hospitals proliferated in the eighteenth and early nineteenth centuries as the practice of medicine became more scientific and professionalized. These hospitals, which catered primarily to the poor, were not exactly beacons of hope: they were underfunded and overcrowded, dark, dirty, and dangerous. Patients didn't just share rooms—some even shared beds. Infectious disease was rampant, and those who could afford it were generally better off receiving medical care at home.

These were the appalling conditions that ultimately spurred the British nurse Florence Nightingale into action. In 1854, Nightingale traveled to Turkey to tend to British soldiers who had been injured in the Crimean War. She was stationed in a makeshift hospital located in a converted barracks. The building was infested with lice, fleas, and rodents. The water was contaminated and drainage was poor; the floors of the wards were covered in sewage. Basic supplies were missing. Patients were draped in dirty, blood-soaked linens.

Despite opposition from military leaders, Nightingale launched a cleanliness crusade. Under her direction, hospital staff bathed the soldiers and laundered their linens, unclogged pipes and drains, replaced vermin-infested floors, and washed the wards with lime. And though the germ theory of disease hadn't yet taken hold, Nightingale intuited what microbiologists would later confirm: a steady flow of fresh air could slow the spread of contagious diseases. So, to improve ventilation at the hospital, she had operable windows installed and vents added to the roof. Mortality rates plummeted.*

In the years after the war, Nightingale published numerous reports calling for reform in hospital design and operation. She advocated for better hygiene, of course, but also recommended that hospitals provide more space per patient, orient their buildings to maximize sunlight, and prioritize natural ventilation. ("To shut up your patients tight in artificially warmed air is to bake them in a slow oven," she wrote in her *Notes on Hospitals*, first published in 1859.) And she waxed poetic about the importance of windows. "Among kindred effects of light I may mention, from experience, as quite perceptible in promoting recovery, the being able to see out of a window, instead of looking against a dead wall; the bright colours of flowers; the being able to read in bed by the light of a window close to the bed-head," Nightingale wrote. "It is generally said that the effect is upon the mind. Perhaps so; but it is no less so upon the body on that account."

Nightingale endorsed an emerging design concept known as the

---

*Nightingale was not the only person to advocate for change at the hospital, nor can the drop in mortality be entirely attributed to environmental changes; she also changed management and operational practices, improved nutrition, developed more systematic intake procedures, and cracked down on corruption that left caregivers without needed supplies.

pavilion-style hospital, in which long, skinny patient wards extended, like fingers, from a central corridor. Two parallel rows of beds ran down the length of each ward, or pavilion. The side walls were studded with large windows, and the pavilions were separated from one another by generously sized lawns or gardens, increasing cross-ventilation. These pavilion-style facilities provided patients with ready access to fresh air, daylight, and nature, and they became increasingly popular throughout the nineteenth century.

But the design trend didn't last. As germ theory and the concept of antisepsis gained ground, hospitals sealed themselves off from the natural world, relying on antibiotics and chemical disinfection, rather than sunlight and fresh air, to reduce the spread of disease. Over the course of the twentieth century, the development of new medical and building technologies, from X-ray equipment to elevators, spurred further changes in hospital design. By the late 1980s, hospitals in the developed world had become cold, sterile environments designed to optimize staff efficiency rather than foster patient healing.* "The state of health-care architecture was frankly pretty dismal," said David Allison, who directs the graduate program in architecture and health at Clemson University. "The environment was focused around a factory-like model of delivering health care."

Into the breach stepped a researcher named Roger Ulrich.

ROGER ULRICH'S JOURNEY to remake the modern hospital was a long and winding road. It also began with one. As a PhD student in geography at the University of Michigan, Ulrich decided to focus his studies on human spatial behavior, interviewing dozens of Ann Arbor residents about how they selected their routes when driving to a local

---

*Hospitals in low-income nations, on the other hand, may struggle to keep things sterile. Twenty-one percent of health-care facilities in the world's least developed countries lack sanitation services, and just 55 percent have basic water services, according to a 2019 report by the World Health Organization and the United Nations Children's Fund.

shopping center. His subjects all lived in the same subdivision, close to a wide expressway with a speed limit of seventy miles per hour. If they took the expressway, they could be at the shopping center in less than six minutes. But more than half the time, they chose to take a slower route—a curving, hilly parkway lined by thick groves of trees—because it was more scenic.

The finding wasn't shocking, but at the time, it was one of the few studies to provide hard evidence for the value that people placed on natural scenery. "There was a broad sense in the humanities—and to some degree the social science community—that beauty was in the eye of the beholder, something impervious to scientific inquiry," Ulrich told me.

After completing his PhD, Ulrich continued his research at the University of Delaware, where he dove deeper into how outdoor landscapes affected people's moods and emotions. For a study he published in 1979, he showed a series of slides to college students who'd just taken a long exam. Half the students saw slides depicting everyday nature scenes—pictures of trees and fields, for instance—while the other half viewed images of streets, buildings, skylines, and other urban environments. Those who viewed the nature scenes felt happier and less anxious after the slideshow, whereas those who saw the urban images tended to feel worse, reporting higher levels of sadness than they had before viewing the pictures. In the years that followed, Ulrich confirmed and expanded on these findings and started to contemplate their potential application. "Is this of any use?" Ulrich wondered. "Where is a large group of people in our society who are experiencing considerable stress for a period of time? The obvious answer was hospitals."

Ulrich knew that firsthand. He had been a sickly kid, a magnet for *Streptococcus* bacteria. "I had an unfortunate gift for getting strep throat all the time," recalled Ulrich, who grew up in southeastern Michigan. Sometimes the strep triggered nephritis, an inflammation of the kidneys. As a result, he became fairly intimate with America's health-care system. "I was quite tired and had all kinds of hospital and office visits, and they were often in pretty brutal circumstances," he said. "They were sterile and emotionally cold—often modernist and functionally efficient but emotionally unsupportive." He much preferred recuperating in his bed at home, taking great comfort in the towering pine tree that stood outside his window.

As he thought back to that pine tree, an idea began to take shape: he'd find a hospital where some patients had views of the natural world and other patients didn't, and he'd compare how they fared. He traveled up and down the East Coast before he found a two-hundred-bed hospital in Pennsylvania that he thought would be the perfect setting for his study. In one wing of the hospital, the patient rooms were almost identical, except for the view: some looked out onto a small cluster of trees, while others overlooked a brick wall. "It was pretty close to being a natural experiment," Ulrich recalled.

Ulrich analyzed the medical records of forty-six patients who'd had their gallbladders removed at the hospital between 1972 and 1981. Half of the patients had been assigned to recovery rooms that overlooked the trees, while the other half convalesced while gazing out at the wall. "It turned out that there was a big effect on pain," Ulrich said. On average, the patients who had nature views needed fewer doses of narcotics than those who looked out onto the brick wall. They were also discharged from the hospital about a day sooner. The study provided hard evidence that Florence Nightingale had been right—a view of nature could indeed be healing—and that modern hospitals had made a mistake in isolating patients from the natural world.

At the time, health-care architects relied more on instinct than evidence and rarely returned to the hospitals they'd designed to see how well they were working. "It seemed like there was an absence of rigorous research on health-care environments and how they influence clinical outcomes," Ulrich said. "The thought occurred to me, 'No wonder hospitals are badly designed.'"

Ulrich's study, which was published in *Science* in 1984, is frequently cited as the beginning of a new era, the birth of what became known as evidence-based design. The paper landed at an opportune time; two new ideas were about to break through in health care. The first was a commitment to providing "patient-centered care," which put the needs of the sick front and center. The second was the birth of the field of "evidence-based medicine," which held that doctors' treatment decisions should be backed by rigorous research. The concept of evidence-based design came to seem like a natural complement. Doctors took an oath to do no harm—shouldn't health-care architects do the same thing?

In the years since, researchers have discovered numerous ways to improve the hospital environment. Many expanded on Ulrich's initial findings, providing even more proof of the healing power of nature. Nearly any kind of nature, they found, seems to do the trick. In the early 1990s, Ulrich reported that heart-surgery patients who'd been randomly assigned to gaze upon nature images had less postoperative anxiety and required fewer doses of strong painkillers than those who viewed abstract art or no images at all. Other researchers found that patients who looked at a mural of a meadow and listened to nature sounds reported less pain while having bronchoscopies and that nature videos reduced anxiety and pain in burn patients who were having their dressings changed. Indoor plants can be beneficial, too; surgical patients with plants in their rooms have lower blood pressure, report less pain and anxiety, and use less pain medication than those in plant-free rooms.

What makes nature so potent? Ulrich believes that the answer lies in what's known as the biophilia hypothesis. The hypothesis, formulated by the famed entomologist E. O. Wilson, holds that because of how we evolved—out in the rough-and-tumble of nature—we have an innate affinity for the natural world. So natural settings and images catch our eye and engage us, cheering us up and taking our minds off our pain and anxiety. "Nature can be quite effective in distracting people in a non-taxing, non-stressful, restorative way," Ulrich explained.

Even a quick dose of nature can trigger significant changes in the immune system. In a series of studies, researchers at Tokyo's Nippon Medical School discovered that walks in the woods could boost the activity and number of natural killer cells, a type of white blood cell that helps vanquish viruses and tumors. The lesson isn't so different from the one that microbiologists have learned: our bodies seem to work best when we stay connected to the natural world—and to the rich assortment of organisms that have surrounded us for much of human history. A healthy indoor environment is one that helps us maintain these links to the wider world outdoors.

Beyond verdant views, hospital windows let the sun shine in, and patients in sunny rooms tend to fare better than those in shady ones. They use fewer painkillers, report less stress, are discharged sooner, and even have lower mortality rates, scientists have found. Though it's difficult to pinpoint the precise mechanism, sunlight can reduce blood pressure, enhance mood, boost vitamin D production, and, we now know, kill pathogens.

It also keeps our circadian rhythms in sync. Our bodies run on daily cycles, and our respiration rate, blood pressure, hormone levels, and immune activity all fluctuate as day turns into night turns into day again. Exposure to ample morning light is what keeps our internal clocks set to the right time, and patients confined to dark, gloomy hospital rooms may find their bodies thrown out of whack. In addition to being too dark during the day, hospitals can fail by being too bright at night; wards that keep the lights on in the wee hours can disrupt patients' sleep, compromising their immune function and delaying their healing.

Some hospitals have been experimenting with "circadian lighting," using artificial light that mimics the way that the intensity and color of sunlight change throughout the day. That means that the lights are cool and bright—rich in blue wavelengths—in the morning, gradually growing dimmer and warmer—more toward the amber end of the spectrum—as evening approaches. (The light-sensitive retinal cells that help regulate our circadian rhythms are most sensitive to short-wavelength light, toward the blue end of the color spectrum. It's this cool, bluish light that triggers these cells to fire off messages to the brain, alerting it that morning has arrived.)

If hospitals really want to help their patients rest, they've got to get a handle on noise. I've had a handful of minor surgeries, and aside from the actual physical pain, the worst part of the experience was the never-ending din. Alarms blared and monitors beeped; the sounds of transport carts and staff conversations echoed down the halls. Hospitals can be as noisy as highways and are often considerably louder than the World Health Organization recommends.

In 2002, Ulrich and his colleagues followed ninety-four patients in the coronary intensive care unit at Huddinge University Hospital in Sweden. Partway through the two-month study, the hospital swapped out the plaster, sound-reflecting ceiling tiles for sound-absorbing tiles. After this switch, the sickest patients seemed to sleep better and display less physiological stress. Even more tantalizing, patients treated when the sound-absorbing tiles were installed were significantly less likely to be readmitted to the hospital within the next three months. (These changes benefited nurses, too, who reported feeling less workplace pressure and strain when the sound-absorbing ceiling tiles were installed.)

Swapping out the ceiling tiles is a good quick fix, but an even better way to give patients peace and quiet is to give them their own rooms. "In the sixties and seventies, the private room was considered something that was nicer, you should pay more for it, and so on," said Kirk Hamilton, the associate director of the Center for Health Systems and Design at Texas A&M University. But a room of one's own is more than mere luxury. "There is a clinical medical reason why patients ought to be in more private settings," Hamilton told me.

In addition to reducing infections, single rooms are quieter, more accommodating to visitors, and better for patient-doctor communication. (One study suggests that emergency room patients who are treated in private rooms, with solid walls and doors, may be less likely to withhold parts of their medical histories and to refuse portions of their exams than those who have nothing more than a flimsy curtain cordoning them off from the crowd.) Hospitals can further improve safety by ensuring that patients are able to remain in the same room from admission through discharge, even if their health deteriorates. Called "acuity-adaptable" rooms, these innovations can reduce the treatment delays, patient falls, and medical errors that sometimes occur when patients are transferred to new units or care teams.

Many of these features are good for the bottom line as well. In 2004, a group of health-care architects, researchers, and executives conjured up a "Fable Hospital," a three-hundred-bed medical center that incorporated design features known to improve patient outcomes and staff satisfaction. "I invented an ideal client," said Derek Parker, an architect who cofounded the Center for Health Design and came up with the idea for the Fable Hospital. "What if we take all this research and put it together in one wonderful place?"

Parker and his colleagues envisioned a hospital with large, private, acuity-adaptable rooms with plentiful daylight; sound-absorbing floors and ceilings; meditation rooms; outdoor gardens; indoor planters; music; artwork; and other amenities. These design features, they estimated, would add about $12 million to construction costs. But—by reducing hospital-acquired infections, patient falls, room transfers, drug costs, and nurse turnover—they'd save the Fable Hospital $11.4 million over its first year of operation, nearly paying for its top-of-the-line design features. "The return on investment was really very compelling," Parker told me.

The Fable Hospital didn't remain imaginary. In 2004, the same year that Parker published an article on his dream hospital, Ohio-Health, a nonprofit health-care system, began planning a real one. It would be a small community hospital in Dublin, a suburb of Columbus, and the executives at OhioHealth were determined to create a state-of-the-art facility that was good for patients and staff alike. The architects they hired gave them a crash course in evidence-based design. Cheryl Herbert, a registered nurse who became the first president of Dublin Methodist Hospital, remembers poring over the literature the architects handed her. "I read and read and reread the Fable Hospital article," Herbert told me. "In the end, we incorporated probably 90 percent of the design elements mentioned in the Fable Hospital into Dublin Methodist."

The hospital, which opened in 2008, doubles down on nature. Leafy green trees stand in the main lobby, a soaring glass atrium that includes a three-and-a-half-story waterfall. Large nature photographs hang throughout the building; all the inpatient rooms and public spaces, as well as many corridors and staff spaces, have windows to the outside. The patient rooms, even those in the ER, are private, and many are acuity-adaptable. Each room includes a family zone, complete with a mini-fridge and a double sleeper sofa where visitors can spend the night. Sound-absorbing ceiling tiles minimize noise. "Quiet is a huge healer of the soul," said Herbert.

In the first few years the hospital was open, patient satisfaction ratings were sky-high—in the ninetieth percentile or higher—while patient falls, hospital-acquired infections, and medication mistakes were rare, Herbert reported in a 2011 article. "The hospital is performing extraordinarily well and has throughout its relatively young life to this point," she told me.

Although the individual design elements that Dublin Methodist incorporated are backed by evidence, hospitals are complex environments, and it's difficult to quantify precisely how much the hospital's design has contributed to its success. "What we learned after we opened was that it's really hard to isolate the impact of the design from the impact of lots of other things," Herbert said.

Indeed, there's woefully little research that tracks the effects of evidence-based hospital design over the long term. "The challenge is that life gets in the way," said Ellen Taylor, vice president for research at the Center for Health Design. "Many times there's just not funding available to continue research once a facility opens."

Even without this long-term data, evidence-based design has changed the look and feel of our health facilities. "It's had impact, no question, on virtually any large hospital today," Ulrich said. Design guidelines issued by the American Institute of Architects call for single patient rooms in all new hospitals. Large windows, courtyards, and atriums are common; meditation rooms, healing gardens, and indoor planters are not unusual. (Some hospitals even offer gardening-focused "horticultural therapy.")

For obvious reasons, most of the research to date has focused on patients—their rooms, their experiences, and their satisfaction. We know far less about how hospital design influences those who are delivering care. This is the next frontier of evidence-based health-care architecture. Researchers are beginning to pay more attention to how medical teams behave, interact, and make medical decisions, and they're thinking beyond the typical inpatient room, venturing into new hospital zones and spaces. Like the high-risk, high-stakes world of the operating room.

OVER THE LAST several centuries, operating rooms evolved in many of the same ways that hospitals did, transforming from large, open theaters in which surgery was a spectacle into sterile, hermetically sealed spaces. As surgical practices and technology advanced, ORs got more crowded and complex. "The operating room is a place of intense human activity and potentially life-threatening and -altering events," said David Allison, the Clemson professor of health-care architecture. "It's a pretty machine-oriented, technologically driven environment. It hasn't historically focused on human needs."

Today's ORs are loud and chaotic, with surgeons, anesthesiologists, nurses, technicians, and students all toiling away at once. Their work is time-sensitive and fast-paced, and over the course of a single procedure, surgical staff perform a staggering range of duties, including retrieving supplies, calibrating equipment, repositioning lights and monitors, tracking vital signs, updating patient charts, labeling and storing specimens, answering phone calls and pages, and managing frequent interruptions and distractions.

Operating rooms are also perilous places for patients; serious complications occur in 3 to 22 percent of inpatient surgeries in developed nations, and researchers estimate that half of these mishaps and misfires are preventable. "In the operating room, the patient is vulnerable, the doctors are doing time-critical, mission-critical stuff," said Anjali Joseph, who directs the Center for Health Facilities Design and Testing at Clemson University. "It continues to be an area where there's need for focus on safer design."

Joseph is leading a large South Carolina team that's trying to create a safer, more human-centered operating room. The four-year project, which began in 2015, involves more than a dozen researchers and clinicians from Clemson University and the Medical University of South Carolina (MUSC). The interdisciplinary team, which includes David Allison, has been driven, in part, by a rare opportunity to put its ideas directly into practice. When the project launched, MUSC was preparing to open two new outpatient surgical centers in Charleston, and the researchers' findings would inform their design.

In January 2018, the team organized a daylong workshop to present their work to a group of health-care design experts. I flew down to Charleston to observe. It was an unseasonably cold morning, and the sun had barely risen when I arrived at the Clemson Design Center, which is tucked into a brick building that used to be a cigar factory. I was tired and uncaffeinated, but Joseph, clad in a bright purple blouse and black blazer, was already wearing her megawatt smile and talking at a rapid-fire pace. She stood before the roughly one hundred attendees and began to share what she and her colleagues had been up to.

Although everyone on the team knew their way around a hospital, they decided to launch the project by examining the operating room with fresh eyes. To do so, they recorded dozens of surgeries in three of the existing ORs at MUSC, tracking the movements and activities of each doctor and nurse. They watched carefully for hiccups and mishaps, moments when the clinicians got distracted, or dropped an instrument, or discovered that a medication they needed was missing. These "flow disruptions" can snowball into major problems, causing delays, jeopardizing staff safety, and raising the risk of medical errors.

The researchers found that flow disruptions were common, counting more than 2,500 across 28 surgeries. Most were minor—a moment when a nurse just glanced down at a pager, for instance—but others required the surgical team to stop their work or repeat a task. More than half of the disruptions were caused by problems with the room's layout—because a nurse needed to walk around a poorly placed instrument table or a surgeon's view was obstructed by a piece of equipment. Disruptions were especially common in the anesthesiologist's work zone, which is crammed with a lot of equipment, and in areas near the surgical table that tend to have a lot of foot traffic.

The findings point to some fairly straightforward ways to improve OR design, such as ensuring that there's plenty of open space around the surgical table and that only the most essential equipment and supplies are stored inside the room. They also suggest that hospital designers should think carefully about the needs of circulating nurses, who were involved in a large proportion of flow disruptions. Circulating nurses are the most mobile members of a surgical team, ping-ponging all around the OR as they monitor their patients and gather supplies, equipment, and instruments for their colleagues. Placing supply cabinets right next to their workstations could curb foot traffic and minimize flow disruptions. (It sounds like common sense, but it isn't a given; in all three operating rooms the researchers observed, the supplies were stored on the opposite side of the room from the circulating nurse's work area.)*

At the design workshop, Allison explained how he and his students had worked to translate these findings into an actual OR prototype, using rolls of colored tape to map out floor plans on the gray carpet of the design center. They considered several OR sizes, testing their hunches using a computer model. Kevin Taaffe, an industrial engineer at Clemson, uploaded data on the movements of every staff member during each of the surgeries they'd recorded, along with maps of the corresponding ORs. Once they had this basic model, which simply replayed the procedures they'd already observed, the researchers could tweak it—moving the door three feet to the left or sending the OR table to the opposite side of the room—to see how the adjustment altered the staff's travel paths. Then they tracked how often each member of the surgical team came into close contact with a colleague.

The model made it obvious that there was a sweet spot when it came to OR size and helped the designers settle on a rectangular room, longer than it was wide, of about 570 square feet. If they made the OR much smaller than that, the number of contacts between members of the surgical staff shot up. Making it larger, on the other hand, increased travel distances without reducing the number of contacts much further. "There's diminishing returns," Taaffe told me.

The researchers also used the model to test an unconventional idea about the position and orientation of the operating table. Initially, they'd placed the table vertically, smack dab in the center of the room, with its head facing the far wall. That was very much in keeping with traditional OR design. "The position of the OR table as centrally located in both axes of the room is something that's frankly never been

---

*Moreover, limiting movement might reduce the risk that clinicians contaminate the sterile zone around the OR table. When the Clemson team put petri dishes in four operating rooms, they confirmed that the high-traffic areas accumulated a lot more microbes than the low-traffic ones.

questioned before and is basically the standard in the industry," Allison said.

In consultation with the MUSC clinicians, Allison and his students came up with a different idea. What if they moved the table out of the center of the room, closer to the upper left corner, and angled it so that it sat on the room's diagonal, with the head of the table facing the corner? This positioning would keep the table, and the patient, away from the door, which they'd placed on the right side of the room. It would open up space near the entrance and at the foot of the OR table, areas that had a lot of congestion. And it would allow the anesthesiologist, who typically sits at the head of the OR table, to set up shop in a protected corner of the room, where she was less likely to be jostled, distracted, or disturbed.

They placed the circulating nurse's work area in the bottom right corner of the room, giving her easy access to the door and the neighboring supply cabinet, and put her workstation on wheels, so she could move it to wherever was most convenient. They called for recessed wall cabinets to reduce the number of nooks and crannies in which bacteria could settle and added a few unconventional perks, such as large floor-to-ceiling windows that would allow OR staff to at least occasionally see the sun. "In the U.S., many clinicians go in to work in the wintertime, never see the light of day, come out of work at night," Allison said. "We don't think that's a healthy environment. We think that's a stress-producing environment."

And in order to make it easy for every member of the surgical team to keep tabs on the procedure and the patient, the designers created large digital displays to be mounted along the walls. The screens broadcast the patient's medical information and vital signs in real time, as well as a live video feed of the procedure. "No matter where you are, what you're doing in the OR, you have access to the information that you need digitally," Allison told me.

The team mocked up the OR in virtual reality, inviting clinicians to provide feedback, and then built a life-size, high-fidelity prototype, stocked with real medical equipment, at the Clemson Design Center. A few hours after the workshop concluded, Joseph, Allison, and their colleagues hosted an official unveiling.

Guests started arriving promptly at 6:00 p.m., decked out in dark suits and lace dresses and leopard-print blazers. They grabbed drinks and canapés—gorgonzola-stuffed meatballs, tiny pimento cheese biscuits slathered with goat cheese and pepper jelly—and wandered to the back of the design center to see the operating room of the future.

"Fred? Fred?" one khaki-clad guest joked, rushing over to the OR table, where a life-size dummy was covered by a white sheet.

"That guy don't look too good," a second man muttered, swilling from his Stella Artois.

Bright white surgical lights, on adjustable, ceiling-mounted booms, shone down on "Fred." There was abundant open space around the surgical table, even with an empty gurney resting along a side wall. The blue-tile floor was clean and clear, free of even the smallest snaking cable. Three entire wall panels had been removed to make room for the windows, and the mobile nursing station was tucked away in the room's bottom right corner. The prototype was clean and minimal and modern. All that was left was to see how well it worked.

WHEN I RETURNED to Charleston three and a half months later, the canapés had been replaced by a bottomless coffee pot, and the design team was all business. A team of nurses rolled into the parking lot just before 8:00 a.m., wearing dark blue scrubs and carrying trays of medical equipment. They were here to put the prototype through its paces.

It didn't take them long to find problems. As two of the nurses stood side by side in the bottom left corner of the OR, opening packages of surgical instruments, they glanced around for a trash can, eventually spotting one at the opposite end of the room. "Is that the only trash can we have?" asked the circulating nurse, a young woman whose dark brown hair was pulled back into a low knot at the nape of her neck.

"Do you think you need more?" said Sara Bayramzadeh, a research assistant professor at Clemson, who was watching attentively from just outside the OR.

"I just need one over there," the nurse said. "And anesthesia will need one."

Bayramzadeh jotted it down.

The circulator and the scrub nurse—a young man sporting a light blue surgical cap—finished setting up the instruments and supplies and then counted them off, noting the exact number of blades, suture kits, and sponges on a large whiteboard. Then the circulating nurse headed to her workstation, signed on to her computer, and entered the relevant information: the room was ready.

The circulating nurse hustled out of the OR to fetch the patient. When she returned, she was pushing a gurney holding a blond child-sized dummy. She moved him onto the operating table, and then the surgeon—or, technically, the pediatric nurse who was playing that role—entered and greeted her colleagues. The circulating nurse wheeled her workstation to the foot of the surgical table and talked the team through the procedure that lay ahead. "We have John Smith," she intoned. "Date of birth: 2/1/2014." They'd be repairing a hernia, she noted, a simple procedure that would take less than thirty minutes and came with a minimal risk of blood loss.

As they went over the details, they didn't even bother to glance up at the wall-mounted displays, which listed the essential presurgical tasks they were working their way through. Bayramzadeh interjected: "Do you think it would be helpful for you to look at the displays?"

The surgeon glanced up at the wall. "Oh, there we go! Right there! Is that it?" The nurses hadn't even noticed them. The displays, mounted above their heads, were simply too high. If they were going to be useful, the nurses said, they really needed to be at eye level.

Then it was time to operate. The nurses brought the instrument stand over to the operating table, gathered towels and drapes, and re-positioned the lights mounted on an overhead boom. The surgeon inquired about turning some music on. "We got any tunes today?" she asked. She prepped the patient, grabbed an imaginary scalpel, and made two invisible incisions in John's abdomen. "Looky there—there it is!" she said. "Hernia!"

She pretended to fix the hernia, then declared victory. The scrub nurse counted the instruments and supplies again to make sure that the team hadn't accidentally left anything inside the patient. "The count's correct," the circulating nurse announced.

The surgeon closed the make-believe wound. "All right, we are done," she said. "See y'all in the next one." The circulating nurse wheeled John out of the OR and into the recovery room. I hoped for a speedy recovery because in a matter of minutes, he was going to be right back in the OR, having his enlarged but nonexistent tonsils out.

Little John's mock hernia repair was just the first procedure in an entire day's worth of simulations the Clemson team had arranged to evaluate how well their prototype worked for surgical nurses. Over the course of eight hours, two teams of nurses ran through an assortment of scenarios, from John's hernia repair and tonsillectomy to a left shoulder arthroscopy on Mr. Smith, an adult-sized dummy.

Not everything went according to plan. During one pediatric tonsillectomy, a nurse dropped the instruments. In the middle of the shoulder surgery, the surgeon's workstation fell off the wall. And the surgeon delighted in inventing emergencies. "Oh no! Something's happened!" she exclaimed during one procedure, jumping back from the patient and throwing her hands into the air. Or "Oh no! I dropped a tube!" Or "Oh no, we don't have enough gas in the room! Our tank just ran out." ("It's like the OR repertory theater in there," one researcher whispered in delight.) The nurses handled it all with aplomb, shaking off the fallen workstation, replacing the dropped supplies, and retrieving new gas tanks.

After each run-through, the researchers descended, pulling the nurses off for one-on-one debriefings. The nurses raved about many aspects of the design. "We usually operate in a closet, so this is super spacious," the scrub nurse reported. The nurses agreed that the supplies were conveniently located and loved the mobile workstation, which allowed them to update a patient's medical record without having to turn their backs on the OR.

But there were also clear areas for improvement. Though the nurses appreciated the uncluttered room, they thought the design team might have taken the minimalism too far. In addition to more trash cans, the OR needed several printers—for images, notes, and specimen labels— and the surgeon needed a place for her personal items.

Finding the right solutions to these problems would involve difficult trade-offs. Sure, the design team could install a few printers, but that would add more clutter. And, Allison told me, they'd intentionally put the displays up high—the lower they are, the more likely they are to be blocked by something else in the OR. "Surgical environments are complex by nature and therefore have a lot of competing demands on the design," Allison said.

In the following months, the Clemson-MUSC team conducted another series of mock operations with a full surgical team working shoulder to shoulder. When the first of the new MUSC buildings opened in the spring of 2019, the operating rooms incorporated many of the researchers' design ideas. Joseph's team is monitoring how well the new ORs are working and developing a "Safe OR Design Tool" that other hospital architects can use to inform their own projects. "We hope to improve the state of surgical environment design in the industry," Allison said.

They're considering applying the same systematic process—of intense observation, design, simulation, and redesign—to other areas of the hospital, ranging from the exam room to the emergency room. If they do, they'll have company. A number of other research teams are producing sophisticated analyses of the inner workings of some of the distinctive spaces that define hospitals. In Philadelphia, for instance, Bon Ku, a doctor of emergency medicine at Thomas Jefferson University, developed a data-logging app that allows researchers armed with iPads to map the intricate dance of activity that unfolds in emergency rooms, tracking the movements and activities of ER staff. One of Ku's goals is to determine whether there are certain aspects of ER design that can encourage clinicians to spend more time talking with patients—and with one another. In one preliminary study, Ku noticed that over the course of a single shift, only 4 percent of one nurse's activities involved interacting with a physician. "Which I think is just crazy," he said.

Ku hopes that his digital mapping tool will help hold health-care designers accountable for the promises they make and bring more analytical rigor to assessments of indoor spaces. Traditionally, when architects or researchers want to determine how well a building is working, they conduct a "post-occupancy evaluation" by surveying or interviewing its occupants. These subjective, qualitative impressions are valuable, but Ku believes that pushing the field of evidence-based design forward will require more quantitative assessment tools and technologies. "The science of the built environment should be as rigorous as the science that we apply to develop new medications," he told me.

While the stakes of design decisions may be especially high in hospitals, the lessons that Ulrich and his peers have learned are widely applicable; plentiful daylight and natural views are restorative no matter where we are. And as the field has matured, the idea of what it means to design for health has expanded. It's all well and good to help patients recover, but wouldn't it be better if we could create spaces that keep people out of the hospital in the first place?

THREE

STAIR MASTERS

I N THE MID-NINETEENTH CENTURY, New York City was a veritable death trap. Yellow fever plagued the ports. Cholera soaked the streets. Tuberculosis tore through tenement buildings. By 1863, New York's annual mortality rate had climbed to one in every thirty-five residents, higher than in any other large city in the United States. In some years, there were nearly twice as many deaths as births.

Local civic groups became alarmed by the mounting casualties— and convinced that they were linked to the city's absolute squalor. At the time, many buildings lacked waste or drainage pipes, so residents tossed refuse directly into the streets. Outhouses overflowed, sending sewage down narrow alleys, and the sewers, where they existed at all, were often clogged. Stables and slaughterhouses were located in densely populated neighborhoods, and farm animals were routinely marched down residential streets, adding their own effluvia to the mix.

Things weren't any better inside the tenement houses, which were dark, dirty, damp, and very, very crowded, providing as little as ten square feet of floor space per resident. Layers of grime caked the walls. Ventilation was practically nonexistent; many rooms had no windows at all, and buildings were often crammed together on small lots, blocking fresh air and natural light. Toilets were typically located in small outdoor courtyards, and their contents sometimes oozed through the walls of neighboring apartments.

No wonder New Yorkers were sick. The sewage that sloshed through the city carried the bacteria that cause cholera and typhoid; the standing pools of water were the perfect breeding ground for the mosquitoes that transmit yellow fever; and the cramped, unventilated apartments helped airborne pathogens spread.

Though sanitary reformers didn't yet understand the biological basis of all of these diseases, they saw a clear connection between the city's filth and its poor health. In 1865, the Citizens' Association of New York, a local civic group, estimated that the city could save as many as ten thousand lives annually if it just cleaned up its act. In fits and starts, the city did. In 1866, it created the Metropolitan Board of Health, which required that yards, outhouses, and vacant lots be cleaned; sent sanitary inspectors to decontaminate the homes of cholera patients; and declared that pigs and goats could no longer roam freely through the streets. Fifteen years later, it established the Department of Street Cleaning, whose workers, clad in all-white uniforms, began sweeping the streets and collecting garbage. (These workers were so effective that the city threw them a parade.)

Legislators passed a series of housing reforms, mandating that every apartment have access to indoor toilets and running water and that every room have a window to the outdoors. (Florence Nightingale would have been proud.) They outlawed certain kinds of cellar apartments, set minimum sizes for yards and courtyards, and restricted rear tenement buildings, which were commonly squeezed behind other tenement houses, on the same narrow lots. The city invested in infrastructure, building more sewers, and, ultimately, a subway, which helped relieve overcrowding by allowing New Yorkers to expand into new neighborhoods.

New York City's mortality rate plummeted, as did the share of deaths due to infectious disease. Although medical advances such as vaccines and antibiotics ultimately helped conquer many diseases, mortality rates fell before these innovations were widespread. As David Burney, an architect who recently served as the commissioner of New York City's Department of Design and Construction, told me, "In the nineteenth century and the early twentieth century, a lot of the solutions to those infectious disease problems were not medical so much as built environment changes. For example, tuberculosis had a lot to do with light and air and planning regulations and density. All those things were driven by the way the city was building."*

As a New York City resident, I'm a direct beneficiary of these reforms. Never once, in my nearly fifteen years of Brooklyn living, have I worried about getting enough fresh air in my home or sloshing through sewage-covered streets. (Though I do confess to griping, from time to time, about uncollected trash.) On my long list of medical worries, tuberculosis and cholera don't even rank; over the course of my lifetime, I'm exponentially more likely to get diabetes or heart disease. Here in the United States, as well as in other high-income, developed nations around the world, chronic disease has replaced infectious disease as

---

*Housing quality and urban planning—or the lack thereof—remain major contributors to infectious disease in many of the world's megacities. Slums and shantytowns tend to be overcrowded, with densely packed, poorly ventilated homes, and many residents have to make do without basic sanitary infrastructure, like sewers to carry away waste and pipes to deliver clean water. Although housing and infrastructure improvements are difficult and expensive, they can yield major health benefits; after the city of Salvador, Brazil, expanded its sewer system, the prevalence of childhood diarrhea dropped by more than 20 percent.

the biggest public health threat. Can architecture once again be part of the solution?

WHEN MAYOR MICHAEL BLOOMBERG took control of New York City in 2002, he was determined to help New Yorkers develop healthier habits. His administration began by trying to snuff out cigarettes, launching a flurry of initiatives that led to dramatic drops in the city's smoking rates. Then officials turned their attention to what they viewed as another pressing set of potential health risks: physical inactivity, poor diets, and obesity.

Like many Americans, New Yorkers were largely sedentary, and their meals were unbalanced; they ate too much sugar, salt, and trans fat and not enough fruits and vegetables. And in New York, as in much of the industrialized world, obesity and diabetes were on the rise. By 2004, nearly one in four of the city's adults was obese, and one in ten had diabetes.

Although it can be tricky to untangle the effects of obesity, inactivity, and poor eating habits, all three can raise the risk of illness. It is of course possible to be overweight and fit, and evidence suggests that body size is less important for health than body composition, or the relative proportion and distribution of muscle and fat. But in general, on a population level, being overweight or obese increases the odds of a variety of diseases, including diabetes, high blood pressure, heart disease, and some cancers. And no matter your size, upping your activity level and jettisoning junk food can have real health benefits.

In an effort to help New Yorkers improve their diets, the city enacted a broad array of new policies, including banning trans fats and requiring chain restaurants to display calorie counts.* It also paid

---

*Some of these moves had more of an impact than others. The trans fat ban, for instance, seemed fruitful; after it took effect, the levels of trans fats in city residents' blood fell by nearly 60 percent. Researchers have also found that trans fat bans are associated with a decline in hospitalizations for heart attacks and strokes. Posting calorie counts on menus is less effective, studies suggest.

particular attention to low-income areas, where obesity, diabetes, and high blood pressure were especially common. Though many of these neighborhoods were thronged with fast-food outlets and corner stores, they tended to lack high-quality grocery stores. So the city doled out incentives to lure supermarkets to these underserved areas and created the "Green Cart" program, issuing permits to hundreds of street vendors who agreed to sell fresh produce.

But food was just half of the equation. Officials also wanted to get New Yorkers moving more, so New York City health commissioner Tom Frieden went to see David Burney, who became the commissioner of the city's Department of Design and Construction in 2004. "He came to me and said, 'You've got to help me with the sedentary lifestyle stuff,'" Burney recalled. "So of course I was like, 'Go away, leave me alone, I'm busy. This is not my problem.' And he was like, 'No, no—you're an architect. You're part of the problem.'"

Although the rising rates of obesity had many causes, Frieden helped Burney realize that the built environment played a role, too. "Architects and planners really have caused a lot of this problem with our drive-in everything and elevators and escalators," Burney explained. "We've made mobility almost unnecessary."

It's innovations like the elevator that made modern New York possible. The city's density depends on being able to build upward—and on having a reliable form of vertical transportation. So perhaps it's unsurprising that in most high-rises, elevators get pride of place; they are prominently positioned in bright, gleaming lobbies, practically demanding to be ridden. Stairwells, on the other hand, are often narrow, dark, and dingy, not to mention hidden away behind heavy fire doors.

We've designed movement out of our neighborhoods, too. For much of the twentieth century, architects and planners gave primacy to the car. As affordable, mass-produced vehicles started rolling off assembly lines, putting car ownership within reach for middle-class families, narrow city streets transformed into asphalt highways designed to speed drivers to their destinations. These highways carved up neighborhoods and facilitated the growth of suburbs. Americans moved out of the urban core; cities began to sprawl.

Compared to compact urban areas, these sprawling neighborhoods have relatively low population densities, with homes that are segregated from local businesses. Rather than featuring compact, interconnected grids of streets, they typically have long, looping roads and cul-de-sacs, with long blocks and few intersections. These design customs tend to discourage walking and make people more reliant on their cars—even for short jaunts to the grocery store, the pharmacy, or a nearby coffee shop.

These pedestrian-unfriendly features can have real health consequences. Adults who live in sprawling areas walk less, weigh more, and are more likely to have high blood pressure than those who live in denser districts, according to a study of more than 200,000 adults living in 448 U.S. counties. The connection turns up again and again. Scientists have found that people are more likely to walk (and bike) when they live in high-density, mixed-use neighborhoods with high street connectivity and that those who live in "walkable" neighborhoods, in turn, have lower blood pressure and are able to better control their diabetes.

These associations hold up in urban areas. New York City residents who live in the most pedestrian-friendly neighborhoods have a lower average body mass index (BMI) than those who live in other parts of the city, even after researchers control for socioeconomic differences. And low-income neighborhoods often lack access to parks, bike lanes, and recreational facilities, public amenities that are associated with higher levels of physical activity.

Over the last two decades, however, some scientists, architects, and planners have been thinking about flipping the script—by designing spaces that help people eat better and move more. Incorporating even a little bit of movement into our daily routines can pay big dividends. In large, longitudinal studies, for instance, researchers have observed that women who walk just ten blocks a day reduce their risk of cardiovascular disease, while men who climb twenty flights of stairs a week—or fewer than three flights per day—face lower odds of an early death.

In New York, officials began to realize that if they could figure out how to engineer environments that encouraged people to exert themselves just a tad more, they might be able to improve public health without issuing a single gym membership. As Burney put it: "We can't make people go to the gym every day, but how can we make people less sedentary without even thinking about it, just in their normal everyday life?"

In 2006, the city's Department of Health and Mental Hygiene partnered with the local chapter of the American Institute of Architects to host its first FitCity conference, which became an annual event. It appointed a Director of Active Design and spent years developing a list of evidence-based active design guidelines.

The guidelines, published in 2010, suggest widening sidewalks and adding protected bike lanes; creating new parks, playgrounds, and other spaces for public recreation; ensuring that all neighborhoods have quality supermarkets; and designing buildings that promote movement. The goal is not to tell people how to live their lives but to make certain healthy behaviors—biking to work, taking a stroll at lunch, grabbing an apple for an afternoon snack—both easier and more appealing. "Design can be used to entice people," said Joanna Frank, who joined the Bloomberg administration in 2010 and went on to become the city's director of active design. Picturesque paths can attract pedestrians the same way that scenic roads attract drivers. "You're much more likely to walk down a street that has trees, that's at a certain scale, that has lots of visual interest," Frank explained.

The guidelines also urge architects to embrace the power of stairs, highlighting studies that have revealed that we're more likely to forgo elevators when staircases are visible, convenient, wide, aesthetically appealing, and architecturally distinct. Installing signs encouraging stair use and displaying artwork or playing music in stairwells can increase stair climbing as well. In 2007, the Geneva University Hospitals, in Switzerland, launched a three-month stair-climbing campaign, hanging posters and stickers encouraging stair use on each of the hospital's twelve floors. Before it began, hospital employees were climbing fewer than five stories a day, on average; during the campaign, that figure soared to nearly twenty-one stories daily. After twelve weeks, the employees had lost weight and fat, their waists had shrunk, their blood pressure and cholesterol had fallen, and their cardiovascular fitness had improved. "It's not rocket science," Frank said. "And that's why it's exciting. It's very achievable."*

Frank and her colleagues have been inspired by the lessons of the past—and the sanitary and housing reforms that helped New York vanquish infectious disease more than a century earlier. "One of the reasons that I was convinced of this being a viable approach was that historical precedent," Frank told me. What makes active design so promising is that, like those earlier reforms, it's a holistic approach that addresses some of the universal, structural causes of disease. "This isn't about one person and their choice of what they eat or how much exercise they get every day," she said. "It's about providing an environment that supports health."

In 2013, Bloomberg issued an executive order requiring city agencies to incorporate the principles of active design when constructing new buildings or renovating old ones and established the nonprofit Center for Active Design. The center, which Frank now runs, hosts the FitCity conferences, organizes an annual design competition, publishes

---

*Architects need to balance these active design principles with ensuring that their buildings remain accessible to people with disabilities. Some strategies, like mounting stair prompts on the wall and playing music in the stairwell, are much better approaches, from an accessibility standpoint, than physically making the elevators harder to reach or moving them farther from the entrance. And if designers and urban planners truly want to ensure that active design benefits everyone, they must find also ways to make public transit and recreational amenities—including parks, paths, playgrounds, gyms, and sports facilities—welcoming to people of all abilities.

checklists and toolkits, and administers Fitwel, a building rating system and certification program that focuses on occupant health.*

The Center for Active Design, which frequently works with affordable housing developers, has made a real push to bring active design into low-income neighborhoods. "We see this very much as an equity issue," Frank told me as we chatted in the organization's headquarters in the bustling Union Square area of Manhattan. "We believe that everybody has the right to have access to a health-promoting environment."

One model of what's possible sits ten subway stops north, in a Bronx neighborhood where nearly 40 percent of residents live below the federal poverty line. That's where the Blue Sea Development Company built Arbor House, an eight-story, 124-unit affordable housing complex. The building, which opened in 2013, includes an indoor gym, complete with a children's climbing wall; a verdant outdoor fitness area; and a ten-thousand-square-foot farm on the roof. The stairs are prominent, wide, and well lit; music plays in the stairwell, and art hangs on the walls. Motivational signs encourage residents to take the stairs instead of the elevators, which have been deliberately slowed to discourage their use.†

The residents have noticed: in a focus group, they told researchers from the Icahn School of Medicine at Mount Sinai that they often took the stairs. One woman said that her children, who loved the music that played in the stairwells, encouraged her to make the climb more often. Another confessed that one reason she hadn't used the stairs in her old building was that they had felt unsafe—they were dark and attracted loiterers. But the Arbor House stairwells were open and bright, with

---

*Under the system, which was created by the Centers for Disease Control and Prevention and General Services Administration, buildings earn points for incorporating active design features, including appealing and accessible stairwells, standing or treadmill desks, exercise rooms, secure bicycle storage, ready access to fresh water, outdoor walking trails, and pedestrian paths to mass transit. Those that score highly enough earn official Fitwel certifications.
†Making elevators slower may curb their use, but this is the kind of design strategy that can make life harder for people who use wheelchairs or have other disabilities that make mobility a challenge. So can "skip stop" elevators, which only stop on some building floors.

glass doors and clear views into the hallways, which made her feel secure enough to skip the elevator.

The Mount Sinai team tracked nineteen Arbor House residents during their first year in the building. When the residents first signed their leases, 79 percent of them reported that they had not climbed a single flight of stairs during the previous week. A year later, that figure had dropped to 26 percent, and there was a modest increase in the number of women who were meeting the CDC's recommendations that adults get at least 150 minutes of moderate exercise a week. "There was a big uptick," said Elizabeth Garland, a doctor and public health researcher who led the study. "People took the stairs—they realized that the apartment building was 'trying to make them healthy.'"*

Active design is no longer niche. The Center for Active Design has trained thousands of designers, developers, and planners, and the Robert Wood Johnson Foundation, a major public health philanthropy and an early leader in the field, has given "active living by design" grants to dozens of American cities, including Nashville, Orlando, Cleveland, Omaha, Seattle, and Honolulu. Internationally, cities from Bogotá to Bristol to Barcelona have begun wresting control of city streets away from cars and adding hundreds of miles of bike lanes as well as new parks, promenades, and pedestrian plazas. The private sector has embraced these ideas, too, with companies from Google to Blue Cross Blue Shield incorporating active design strategies into their corporate campuses.

But if we really want to change health habits, the best time to start is in childhood, and some of the most innovative active design projects focus on our youngest citizens. Kids spend as many as half their waking hours in schools and are sedentary for most of that time; a few small design tweaks could potentially improve the health of hundreds of children in one fell swoop. Terry Huang has been thinking about this possibility since 2005, when he began a five-year stint overseeing childhood

---

*Some of these design strategies can support mental health as well. Walking paths, parks, sports facilities, and big, open, central staircases are all spaces where people can interact, socialize, and build a sense of community.

obesity research at the National Institute of Child Health and Human Development. As he watched the field of active design begin to blossom, he started to contemplate how its principles could be applied to schools. He laid out his ideas in a 2007 paper, imagining schools with enticing staircases and standing desks, vegetable gardens and juice bars, teaching kitchens and food demonstration areas, and a wealth of spaces in which kids could move their bodies, from Frisbee fields to dance studios. "At the time it was kind of a fanciful idea," said Huang, now a professor at the City University of New York. But then he heard about two schools that were being planned in rural Virginia.

BY 2009, DILLWYN PRIMARY SCHOOL, in Buckingham County, Virginia, was bursting out of its brick walls. Only half of the school's three hundred students, who ranged from preschoolers to third-graders, fit inside the single-story building, which was more than fifty years old. The rest spent their days in dilapidated trailers located behind the school. "They had little tiny small windows, moldy carpet," said Pennie Allen, who was the school's principal at the time. "They leaked. The handrails were rotted."

The main building was in slightly better shape, but it had no gymnasium; when it rained, the gym teacher would commandeer a corner of the cafeteria for an improvised round of bowling. It didn't have air conditioning either, and on warm days, Allen would run around to all the classrooms, doing temperature checks with a couple of handheld thermometers. If the mercury climbed too high, the school would simply close. "When you go into a room with twenty-five kids and it's ninety-two degrees in there, there's not much learning going on," she told me.

Dillwyn Primary wasn't the only local school in need of new facilities. Buckingham's population, which is about two-thirds white and largely low income, had been growing, and its small, aging schools hadn't kept up. Buckingham is located in the heart of Virginia, stretching across nearly six hundred square miles of rolling green hills. The county bristles with pine and oak forests and sits atop a rich mineral belt; in the nineteenth century, it was dotted with gold mines. Mining remains an important local industry—Dillwyn Primary School was situated between a slate quarry and a kyanite mine—and Buckingham has become known for its shimmering blue-gray slate.

Allen grew up in Buckingham, and she knew the challenges that local kids faced. More than 20 percent of Buckingham children live below the poverty line, and 70 percent of the elementary school students qualify for free or reduced-priced lunch. They tend to struggle on the state's standardized tests. But Allen—a natural nurturer with six grandchildren and a sunny southern drawl—loved working in Buckingham and felt protective of her students. "I've always had this passion—I want the kids in rural Buckingham to have the same opportunities as a kid growing up anywhere," she said.

She worried about them, too. She had studied nursing before training as a teacher, and her father had died young, from heart disease. Despite the county's bucolic setting, kids didn't get much exercise. Many lived too far from school to walk or bike there, and there were few public recreational facilities. The county did have a youth sports league, but many families lacked a reliable means of transportation, or the gas money, to get their children to practices and games.

Allen tried to make what difference she could at Dillwyn. She won federal grants that allowed the school to bring in Zumba instructors, offer a sports-centric after-school program, and buy fresh fruits and vegetables for students to snack on throughout the day. But there was only so much that one driven, dedicated principal could do.

Then Allen got word that the county had decided to create two new schools for Buckingham's youngest learners: a primary school that would run from kindergarten to second grade and an elementary school for third- through fifth-graders. County officials asked Allen to serve as the planning principal for the schools and hired VMDO Architects, a design firm based in Charlottesville, Virginia, to create them.

VMDO had developed a reputation for creating cutting-edge,

eco-friendly schools, and that's what the firm focused on as it started sketching out ideas for Buckingham. Then several team members went to a lunchtime lecture at a local design center. The speaker was Matthew Trowbridge, a pediatrician and public health researcher at the University of Virginia. Trowbridge had recently begun collaborating with Terry Huang, and during his talk that day, he outlined the ways in which architecture and urban design could be used to promote public health.

The VMDO architects were impressed, and they reached out to Trowbridge. "We were really eager and excited to get to know him and his work and potentially partner with him," said Dina Sorensen, a VMDO associate who was one of the lead designers on the Buckingham project. Trowbridge suggested she read Huang's paper.

To Sorensen, it was a revelation. "It was like someone just handing me a pot of gold," she told me. "I'm not the same person today, because of that paper." She had long been animated by the idea that architecture in general, and schools in particular, can unlock human potential, fostering learning, curiosity, and creativity. As Sorensen read about active design, she saw a whole new set of possibilities opening up before her, an opportunity to create schools that were good for children's bodies as well as their minds. "I was totally smitten," she said.

Trowbridge introduced Sorensen and her colleagues to Huang, and together they began to brainstorm about Buckingham, dreaming of two dynamic, kinetic schools that nurtured and nourished children. The interdisciplinary design team wanted to reimagine every nook and cranny, each aspect of the daily routine, creating habits that led to a lifetime of health. When Allen heard the pitch, she found it thrilling. "It wasn't just about [test] scores," she told me. "It was about health and wellness and the whole person."

The county eagerly signed on to the vision, and the planning group got to work. The team decided that rather than constructing the schools from scratch, they'd completely renovate two existing school buildings, which were located next to each other on a forty-acre site.

The idea was to turn the two structures, one of which had been completely abandoned, into neighboring primary and elementary schools, joined by a glass bridge. The schools, which would be named Buckingham Primary School and Buckingham Elementary School, would each have their own principals but share a cafeteria, library, and other common spaces. Together, they would serve all one thousand of the county's kindergartners through fifth-graders.

With these details nailed down, the team started translating the scientific evidence into actual design features. They devised a number of what they called "movement temptations" to appeal to kids' natural desires to run, jump, climb, and explore. In the elementary school lobby, they created a monumental staircase, surrounded by large windows and bathed in natural light. They gave the staircase, and all the others in the building, neon railings and posted playful prompts near each set of steps. ("Hop on up!" one sign says in bright red text. "Get out of your chairs! Jump up! Jump down! And hop on up the stairs!")

They transformed the long, featureless hallways into "learning streets" studded with reading nooks and small group workspaces stocked with soft, colorful, movable furniture. And to enliven the experience of walking, they hid the outlines of animal tracks in the building's terrazzo floors. "So just like if you're on a hike and you're looking around and you see a track, it makes you anticipate the joy of walking," Sorensen said. "It creates this long-standing attachment to place that embeds the idea that walking is good, moving is rewarding."

In the primary school, they created a big, open foyer, called the "Woodland Hub," at what used to be the intersection of two dark corridors. In the center, on a slightly raised wooden platform, they fashioned a kind of abstract playground, called the Tree Canopy, arranging a row of tall boards to loosely resemble a grove of trees. Holes carved into each board allow kids to crawl into, around, and through the simulated forest.

To further reinforce the idea that exercise can be fun, they installed glass panels in the walls of the campus's two gyms—one in the primary school and another in the elementary school—so that students passing by in the halls could see their schoolmates in action. And because they knew that not all students were budding basketball players, they set the elementary school gym aside for organized sports and turned the primary school gym into a soft-floored multipurpose room for free play.

It's a thoughtful touch, one that I would've been grateful for when I was a kid. (I spent my childhood as an avowedly team-sports-averse gymnast of middling talent, and I would've loved to have had a space where I could practice cartwheels or play a game of four square.) The non-sports gym is a nod to the fact that there are lots of different paths to health and fitness and that it's important to meet kids where they are, something the design team took pains to acknowledge in other ways as well. For instance, they ripped a set of old metal lockers out of one of the elementary school hallways and replaced them with a wide, undulating ribbon of natural wood that became a hallway-length bench. "When we think about movement, we also anticipated diversity of need," Sorensen explained in a talk she gave about Buckingham. "And what entices walking for some children and adults is knowing there's an attractive place to rest or pause."

Inside the classroom, however, the designers wanted kids to spend less time resting on their rears. When we sit, our bodies undergo a cascade of physiological changes: muscles slacken, fat burn falls, circulation slows, blood sugar rises, and insulin production spikes. In the long run, spending too much time stationary increases the odds of cardiovascular disease, type 2 diabetes, and cancer.

In a classic study published in 1953, the British epidemiologist Jerry Morris examined the health records of tens of thousands of men working as drivers or conductors on London trams, trolleys, and double-decker buses. The drivers spent their shifts sitting, while the conductors were often on their feet, walking up and down the aisles and stairs to collect tickets from passengers. Coronary heart disease, Morris found, was more common and severe—and developed earlier—

among the sedentary drivers than among the more active conductors. Studies conducted since then suggest that long, uninterrupted sitting sessions appear to be especially dangerous; adults who sit for extended periods without getting up are at a higher risk for early death than those who accumulate a lot of sedentary time over the course of a day but regularly get up to take small breaks.

The concern has inspired furniture that's designed to get us back on our feet, like the increasingly ubiquitous standing desk. The benefits of these desks are still being debated; some studies show that standing desks can reduce the amount of time that office workers spend sitting and help schoolkids burn more calories, but whether they lead to substantial health improvements, especially over the long term, is less clear. Nevertheless, they've become immensely popular, and the Buckingham team wanted to maximize the opportunity for classroom movement, providing more literal wiggle room for students. So they purchased adjustable-height tables, as well as "dynamic furniture," including stools with rounded bases, which rock and roll and pivot in three dimensions, moving with children's bodies as they shift in their seats. "Children naturally fidget anyway, but in the traditional classroom setting, they're constrained," Huang said.

Perhaps the biggest challenge the design team took on was the cafeteria, a space that is traditionally an afterthought in school design. "Our big radical idea was to completely transform the cafeteria into the most important classroom in the school," Sorensen said when she presented the Buckingham project at New York's 2015 FitCity conference. They wanted to turn the food service workers into genuine educators, to create a lunchroom that was more than just a place to offload kids while they scarfed down their tater tots. So they decided to place an open, airy dining commons right in the heart of the campus and to keep the food preparation process visible by eliminating the walls between the kitchen and the dining area. The cafeteria staff were already doing all their baking from scratch—a relatively unusual practice for a public school—so the architects designed windows that allowed students to see into the bakery as they entered the dining commons.

The schools couldn't do hands-on food education in a commercial kitchen, so the designers added a small teaching kitchen and a food lab scaled to the kids' size. They imagined the spaces being used to teach students how to clean, prep, and cook fresh produce or to host visits from local farmers. They even built bookshelves into the lab, which they hoped would be stocked with books about food, farming, and nutrition. "So if you're a kid sitting at lunch and you're like, 'Hey, I have this idea about carrots,' you can run to the little library, grab a book," Sorensen told me. "That was our vision." They set aside a plot of land, right outside the cafeteria, for a kitchen garden, which could be used both as a classroom and to provide fresh produce for school meals. "It sets up this idea that food comes from somewhere," Trowbridge said. (Cooking classes and school gardens can increase kids' food knowledge and their willingness to try, preference for, and consumption of vegetables.)

Sorensen and her colleagues knew that some kids might need an extra push to give leafy greens a go, so they borrowed some ideas from behavioral economists, who have demonstrated that we can steer people toward certain choices by subtly altering the way those choices are presented. As anyone who has ever kept candy on their desk has discovered, we're more likely to consume food when it's within easy reach. Making wholesome food more prominent, accessible, and convenient—or, conversely, making junk food more difficult to acquire—can nudge our diets in a healthier direction.*

Take a 2013 study in which scientists simply repositioned the milk in a California elementary school. Typically, both white and chocolate milk were readily available, stacked in neighboring crates in front of

---

*A clever study of vending machines reveals just how much we value instant gratification. Bradley Appelhans, a health psychologist at Rush University Medical Center in Chicago, engineered a system that forced vending machine customers to wait twenty-five seconds to receive junk food but delivered healthier snacks immediately. When he deployed the system around campus in 2015 and 2016, the proportion of healthy snacks purchased ticked up modestly.

the lunch counter. But for one week, the researchers hid the chocolate cartons behind the counter and posted a sign telling students that if they wanted chocolate milk, all they had to do was ask. The share of students selecting white milk jumped from 30 to 48 percent. Other studies have shown that kids eat more fruit when it's presented in a fun, attractive way—speared on decorative toothpicks and pinned to a watermelon—and college students consume more apple and carrot slices when the produce is sitting right on the table rather than placed two meters away. Ambience matters, too. We don't like spending time in unpleasant, stressful environments, and restaurants with glaring lights or loud noise can prompt us to dine and dash, cramming our meals in as quickly as we can manage, or just plain overeat.

The Buckingham design team shared these lessons with administrators, recommending that the schools make healthy foods easy to see and grab—by putting, say, a basket of fresh fruit right by the cash register. To encourage kids to drink more water and fewer sugary, calorie-laden beverages, they installed a fresh water station next to the milk and juice cart. (Kid-sized fountains also dot the hallways, and every classroom has its own water cooler.) And they created colorful signs, touting the benefits of water consumption and the components of a healthy meal, to hang throughout the schools.

Finally, because children are more active outside than inside, Sorensen and her colleagues thought carefully about how to make the most of the sprawling property, which features a stream and a natural wetland. They created a meandering network of walking paths, which wove around several playgrounds and sports fields, a picnic knoll, a frog bog, newly planted fruit and nut trees, and a "grab-'n-go" berry patch. There's a full-fledged outdoor classroom, the art and music studios have their own outdoor terraces, and each of the kindergarten classrooms opens onto a small outdoor play area. The team hoped that this ambitious landscaping plan would turn the campus into a community park, attracting children and adults from all over Buckingham, rather than a mere school playground. "When you look at a school landscape as a community asset, it's the most underutilized public land we have," Sorensen told me.

The schools represented a major, taxpayer-funded investment, and the design team wanted to make the most of it. So they created a large community commons, with amphitheater seating, in the elementary school lobby and made sure to involve county residents in the design process, convening focus groups, consulting with local NAACP and 4-H chapters, and organizing community garden workshops. "Because it's such a big investment, we really have the community's attention," Trowbridge told me. "Whether they agreed with the exact shape of the stair or whatever, there was a very robust conversation about health and health promotion for kids that occurred because of the school renovation. And I think that's a really cool public health opportunity."

THE SCHOOLS OPENED in the fall of 2012, and the students fell in love on day one. "I just remember them with their eyes just wide open, saying 'We love this place,' and 'Can we spend the night here?'" recalled Pennie Allen, who became the principal of the new primary school. The kids loved the gyms and the reading nooks; the jiggly, just-the-right-size furniture; and the big, bright cafeteria.

For her part, Allen did exactly what the design team had hoped, embracing the new building as an opportunity to create a culture of health. She assembled a nutrition committee and invited a nutritionist to come speak in the community meeting room. She nixed some of the school fundraisers that involved selling junk food and replaced them with 5K runs. She stopped giving cupcakes to kids who made the honor roll and created an annual movement-focused fall festival, where students ran relay races and built scarecrows. She instituted "brain breaks," during which students would put classwork on pause for two minutes while they did a quick round of jumping jacks or played a short, active game. And during the schoolwide announcements each morning, she highlighted the most nutritious options on that day's lunch menu. "I hope you make a healthy choice today," she would say. And the students responded; it was not uncommon for kids to run up to her in the halls and proudly announce what they'd eaten. "She totally rocked it," Sorensen told me. "She understood how to build a culture, a new culture, in small ways. And she just has a phenomenal grasp of what motivates kids."

A year after the schools opened, a research team led by Huang returned to survey staff about how things were going. "We found a lot of inspiration and hope," Huang told me. "Teachers and administrators saw new possibilities, and because of the new possibilities, they introduced new practices on their own." Some teachers started incorporating the garden into their lessons—students grew seedlings and made mini-pizzas with vegetables they harvested from the garden—and the after-school program used the teaching kitchen for salsa- and smoothie-making classes.

Some teachers had been so enthusiastic that they'd launched initiatives to promote employee wellness, selecting "nutrition leaders" to bring wholesome snacks to meetings and launching a weight-loss contest for teachers and administrators. And the efforts yielded results; the share of staff members eating high-fat diets decreased from 74 percent to 57 percent, Huang found.

Other staff members had been more difficult to win over. The food service workers initially grumbled that the vast kitchen required too much walking and fretted that there wasn't enough freezer space. (Allen had to remind them that the whole idea was to serve less frozen and processed food.) And though the design team had wanted to turn the kitchen workers into educators, few of them had been trained in nutrition. "Without a program designed and targeted at the food service staff, they were not going to be able to actually optimize the use of the new kitchen," Huang said. "You have to have the space to make new things possible, but people don't just naturally and organically take advantage of the new space."

Some of the teachers weren't too keen on all the walking either, complaining that it took too long to get from their classrooms to some of the common areas. "Promoting walking in school environments seems like a no-brainer, but there's a lot of resistance to it from the teaching perspective because they're thinking, 'We lose instructional time every time we have to go somewhere,'" Sorensen said. And the dynamic furniture drove some teachers bananas; Allen kept finding the rocking stools stuffed into closets.

The designers had never believed that their buildings would be a cure-all. Instead, they had conceived of them as hardware. The schools' policies and programs—banning bake sales and offering after-school cooking classes, for instance—were the software. Ideally, the software and the hardware would reinforce each other. "When those two are really working together in a sophisticated way, then the likelihood of behavioral change is higher," Sorensen said. But after the Buckingham schools opened, it was clear that the software was a bit buggy.

AND SO, WHEN Sorensen and her colleague Kelly Callahan drove me to Buckingham, they worried a bit about what they'd find. It had been five years since the schools had opened, and they hadn't been back since Pennie Allen had retired at the end of the previous school year. They were curious to see how the schools were holding up without their biggest champion.

We arrived on a picture-perfect spring day. The sky was a bright, clear blue, and pink and white blossoms burst from the trees; a yellow butterfly bopped around the bushes. With some trepidation, we ventured into the main lobby. The sounds of basketballs thumping and sneakers squeaking on the gym floor echoed from down the hall as we made our way into the dining commons.

It looked nothing like my elementary school cafeteria, a claustrophobic, windowless space with cramped picnic-style tables that folded down out of the wall. Here, small white tables, each surrounded by eight wooden chairs, were arranged in tidy rows. The overhead lights gave off a soft, warm glow. Sunlight streamed in through the wall-to-wall windows, which provided a view of the expansive grounds; the fruit trees were in bloom, and a group of primary school students scampered around a distant playground.

Sorensen was pleased by how good it all still looked and delighted to see huge bags of flour stacked in the bakery, an indication that the schools were still doing all their own baking. But then she spotted a PET Dairy–branded freezer, packed with ice cream novelties, sitting in the serving line. "That ice cream was never here when Pennie was here," she said. And she wasn't thrilled by what she found when she peeked inside a refrigerated cooler full of milk cartons. The white milk all sat at the bottom of the cooler, with the chocolate milk stacked on top, where it was highly visible and easy to reach. "They've kind of abandoned simple things you can do," she said as she reached into the cooler and rearranged the milk.

When I walked over to the serving area, I spotted a sign that helpfully broke down the components of a healthy meal, depicting a colorful plate filled with vegetables, whole grains, protein, and fruit. But I couldn't see how students could actually assemble a plate that looked like that, even if they'd wanted to; right above the illustration was the day's lunch menu: corn dogs, french fries, baked beans, frozen juice cups, and a cookie. "Corn dogs—ugh," Sorensen sighed. (Although school districts have control over their own lunch menus, most public schools depend on federal subsidies and commodities to feed their students, and budgets are extremely limited. So while no one's making Buckingham serve corn dogs, they can't exactly go on Whole Foods shopping sprees, either.) On our way out of the cafeteria we discovered that there was a popcorn machine in the Food Lab and that the shelves of the food library were bare.

There were other disappointments upstairs. On the ride down to Buckingham, Sorensen had told me about her favorite space on campus—a cantilevered corner room on the primary school's second floor. The sharp angles and glass walls let students look out over the grounds, as though they're sitting in the prow of a ship. "You literally get this hug from the building," Sorensen said. But when we stepped inside the room, the curtains were drawn across the windows. And what was supposed to be a small group meeting space—with soft, lounge-like seating and cushions—had been transformed into a full-time classroom crammed with rows of desks. Apparently the student population had been growing, and the school needed the extra classroom space.

Sorensen is philosophical about these discoveries. There's not much she can do about a booming student population, and she knows that there's a lot of pressure on the staff to improve literacy and academic outcomes. "I take the long view," she told me. "Their healthy eating campaign or stocking the food library or whatever may just fall to the wayside for a year because they're all rallying around something else that's really impacting a lot of kids. In my view, it's okay. It's okay if spaces shift use, it's okay if one year they get it, and it's fully operational, and the next year the champions change."

Research on the schools' impact has been mixed. Huang and his colleagues found that after a year at the school, students who'd entered as fourth-graders displayed modest but measurable increases in their nutritional knowledge; some students reported that they'd learned these new facts from the educational signs that were mounted around the building. But kids who'd started as fifth-graders didn't score any higher on a nutritional quiz after a year at the school.

When Huang's team used accelerometers to monitor students' daily movements, they discovered that the new schools seemed to make the kids less sedentary than they would have otherwise been. Overall, they concluded, the schools appeared to increase the amount of light physical activity students got by more than an hour a day. "When you compare kids at Buckingham to kids in similar schools that did not have any physical transformation, our kids were way less sedentary than the control kids," Huang said. "I think it provides nice preliminary evidence showing the potential benefit of school design in promoting health and wellness."

It wasn't entirely good news; the schools seemed to reduce the amount of moderate to vigorous physical activity students got by about ten minutes a day. Huang hypothesized that the long walking distances deliberately designed into the schools, especially the relatively lengthy stroll from many classrooms to the outdoor play areas, may have eaten into the time allotted for recess, ultimately reducing the time kids spent in energetic play. The increase in light activity more than made up for this slight decline, but the finding highlights the fact that built environments—and human behaviors—are complex, and that even carefully considered design decisions can have unexpected effects.

Though few have embarked upon redesigns as ambitious as Buckingham, schools around the world are finding ways to incorporate elements of active design. They are reorganizing their lunch lines; mounting stair prompts; planting rooftop gardens; adding springy athletic flooring to their lobbies; and turning underused spaces into fitness centers open to students, teachers, and parents. Amid all this enthusiasm, though, Huang would like to see a more systematic effort to track and evaluate these projects. "That continues to be a source of my frustration," he told me. "There's a lot of talk, a lot of interest, but how are we really growing the evidence base?"

An increasing number of studies show that active design can change behavior—boosting stair use or local cycling rates, for instance—but it's much harder to prove that these changes have real, long-lasting benefits. It's great to design a cafeteria that prompts students to serve themselves more fruit, but are a few extra apple slices a day actually making kids healthier?

Design can be a powerful part of a public health campaign, but making a real dent in chronic disease will require a multifaceted approach that includes restrictions on food marketing, improvements to school lunch guidelines, reforms to the health-care system, and changes to the agricultural policies that have made fats and sugars so cheap and fruits and vegetables so expensive.

As active design becomes more popular, we would do well to heed the lessons of Buckingham, to remember that while buildings can be catalysts for change, the process isn't quick or easy. "It takes a lot of effort, so I feel like what Buckingham has accomplished, even if it looks different when we walked in today than it did five years ago, is remarkable," Sorensen told me during our visit. "For them to take it on is profound."

Teachers and administrators have told Sorensen that a cultural shift is under way and that a new sense of community is slowly developing.* There have been setbacks in Buckingham, but I also saw some real reasons for hope. I saw kindergartners traipsing around their rain gardens and a teacher leading her students around the campus grounds like ducklings as she gave a lesson about deciduous trees. I watched as class after class of students thundered up the stairs and spotted large superhero decals—Batman, Spider-Man, and Captain America—plastered above the main landing, greeting them as they climbed. And I discovered that the Tree Canopy, which is sometimes used for occupational therapy for kids with motor difficulties, is well cared for and beloved.

Cultural change is a long-term commitment, and the Buckingham designers believe that their job is not simply to prompt change but also to anticipate it. "The building allows them to grow and have the confidence they can continue to evolve and not get held back by the design or the built environment," Sorensen explained. After all, whatever the software glitches, the hardware is still in place. If an enterprising

---

*In the fall of 2017, Sorensen moved to a new design firm, the DLR Group in Washington, D.C., but she continues to follow the schools' progress and give talks about the work she did there.

teacher decides she wants to expand students' nutritional knowledge, the food library will be there, shelves waiting to be filled.

Fortunately, as Buckingham finds its way forward, it doesn't have to choose between doing what's good for kids' bodies and what's good for their minds: children who have healthy habits—who eat fruits and vegetables and are physically active—tend to do better in school, numerous studies show. But that's just one potential strategy for boosting students' brainpower, and scientists in labs around the world are investigating other ways to design indoor spaces—from schools to offices— that allow us to perform at our cognitive best.

FOUR

THE CURE FOR THE COMMON CUBICLE

I N THE SPRING OF 2016, eight employees of the Mayo Clinic's medical records department packed up their belongings, powered down their computers, and moved out of their longtime office in the heart of Rochester, Minnesota, and into a brand-new workspace just a few minutes away. There, they made themselves at home—hanging up Walt Disney World calendars, arranging their framed dog photos, and settling back into the daily rhythms of office life.

Then scientists started messing with them. They cranked the thermostat up and then down. They changed the tint of the windows and the color of the overhead lights. They played irritating office sounds through speakers embedded in the ceilings: a ringing phone, the clack of computer keys, a male voice saying "Medical records."

When I visited the office one warm morning in June, the recording was playing on a loop. "I've timed it," Randy Mouchka, one of the relocated employees, told me with exasperation. "It's fifty-five seconds." The air felt stale and stuffy, but the sun was glinting through the windows—an improvement over last week, Mouchka said, when the shades had been pulled down every day.

Down the hall, in a glass-walled control center crammed with computers, the scientists were keeping a close eye on Mouchka and his colleagues. One monitor featured a live video feed; others displayed real-time data on the light levels, air temperature, humidity, and atmospheric pressure, as measured by the one hundred or so sensors scattered around the office. The workers were wired up, too: a large monitor showed the readouts from biometric wristbands that measured their heart-rate variability and the electrical conductance of their skin, both crude measures of stress. The researchers were monitoring all these reactions as they subjected the employees to different office conditions.

These eight medical records employees were the first guinea pigs at the Well Living Lab, an immersive research facility where "Big Brother" meets big data. The lab—a collaboration between the Mayo Clinic and Delos, a New York–based real estate company—was custom built to facilitate multidisciplinary research into how the indoor environment influences human health and well-being. Since the lab opened, its investigators have spent much of their time trying to understand the privations of life in the modern office.

Scientists have spent decades investigating workplaces, from the factory floor to the executive suite, and it's abundantly clear that the physical work environment can affect workers' comfort, stress levels, and performance. Background noise can impair memory, sap motivation, and cause fatigue. Insufficient lighting can spur errors. Frosty and scorching air temperatures can not only cause discomfort but also make tasks feel more difficult. When the air in offices is saturated with pollutants—including carbon dioxide and volatile organic compounds, which are emitted by many common furnishings and products—workers fare poorly on tests of cognitive function. (Thanks to our own exhalations, the levels of carbon dioxide in conference rooms, lecture halls, and classrooms routinely rise high enough to make us drowsy and muddle our thinking, studies suggest.)

The Well Living Lab represents a different scientific approach. The lab gives researchers precise, fine-tuned control over a multitude of environmental variables, and its subjects spend months performing their actual jobs in a space that functions as a real, open-plan office. It's one of a handful of "living labs" that have popped up around the world and part of a renaissance of research into office life. Armed with an array of technologies, from biometric sensors to mobile apps, scientists are developing a more sophisticated, fine-grained understanding of how office design affects human cognition, performance, and behavior—and what employees need to be happy, comfortable, and productive.

DELOS WAS BORN IN 2009 when Paul Scialla, a former partner at Goldman Sachs, decided that he wanted to create a company devoted to "wellness real estate." The company—which offers consulting services and develops health-focused building management systems—made its first big splash in 2014, when it released the WELL Building Standard, a collection of evidence-based guidelines for designing healthy buildings, from using paints that release minimal levels of potentially toxic compounds to organizing cafeterias so that they prominently display fruit and vegetables. Buildings that meet enough of the standards can become "WELL Certified," in much the same way that sustainable, eco-friendly buildings can earn a LEED certification.

But in putting the standard together, Delos had noticed limitations in the literature. A classic study of thermal comfort, for example, might ask a volunteer to spend three hours in a small, windowless "climate chamber," working through math problems while scientists tinker with the temperature. And the researchers might conclude that the volunteers made more mathematical mistakes when the chamber was sweltering and, thus, that heat impedes cognitive performance.

But volunteering to spend a few hours reliving high school algebra in a climate-controlled box is not the same as performing your actual job in an office that feels like an oven. So Delos partnered with the Mayo Clinic to create something that would occupy a scientific sweet spot—more realistic than many laboratories and more controlled than the offices used for field studies.

They sketched out a 7,500-square-foot dream lab. The facility, which cost more than $5 million to build, is endlessly adaptable. The window shades can be programmed to rise and fall at specific times of day. The tint of the windows and the color and intensity of the lights can be adjusted remotely. Speakers installed in the ceiling can play an entire library of sounds—including white noise and muffled conversations—at varying volumes. "We can move walls, we can move plumbing, we can move ducts," Brent Bauer, the lab's medical director, told me. They can transform the lab from a large, open-plan office to a cluster of apartments or hotel rooms, where volunteers might live for weeks or even months.

When I arrived in Rochester, the researchers had just launched their pilot study, designed to validate the lab's technological systems and approach. "To be honest, this is a very preliminary experiment," Bauer warned me. Over the course of eighteen weeks, Bauer and his colleagues were creating environments that they hypothesized would have varying effects on workers' comfort and stress—quiet and temperate one week, warm and noisy or cold and dark another. They were monitoring workers' responses with surveys, interviews, and the biometric wristbands. Even for a relatively simple experiment, the amount of data streaming in was enormous, the lab's director of technology told me as we watched all the assorted measurements flash across the control room monitors.

When the study concluded a few months later, the Well Living Lab investigators began plucking observations out of this river of data. Of all the variables, the subjects found the changes in temperature to be the most noticeable, and they were unhappy and uncomfortable when the space was chilly. "They'd walk up and down the stairs to warm up, they brought in gloves, blankets," Bauer said. They were especially miserable during the weeks when both the thermostat was turned down and the blackout shades were closed—the dark, cold office made it feel like winter, they reported. The participants said that the discomfort made it harder to get work done and that their inability to change the conditions around them left them feeling helpless.

None of this was terribly surprising—who wants to spend eight hours a day in a cold, dark office?—but it did speak to the value of assessing multiple variables in concert. So did another finding: Dissatisfaction with one aspect of the environment could color workers' impressions of its other attributes. For instance, when the office was dark, noisy, or cold, the workers had more complaints about the air quality— even though the researchers didn't adjust the air at all throughout the course of the study. "People's perception of the environment isn't very granular," said Anja Jamrozik, one of the lab's cognitive scientists. "When they're dissatisfied, they have an intuition that things aren't right, and they look around to see what might be affecting them. And air—since it's not visible but it's everywhere—is a thing that people can attribute their problems to."

It's a simple insight, based on a small study, but its implications could be profound. Building managers tend to get a lot of complaints about air quality, which can be tricky to improve, Jamrozik told me. But the lab's findings suggest that in some cases, it could be the temperature, acoustics, or lighting that needs adjusting. Indeed, other teams have turned up evidence that environmental variables can interact in complex ways; in one study, the more that researchers dialed up the background noise in a testing room at the University of Nebraska, the less satisfied people were with the air temperature.

Jamrozik is leveraging her training in cognitive psychology to help the Well Living Lab team dive deeper into the minds of office workers. She worked with her colleagues to develop the "Think It Out" app, which measures what's known as executive function, a set of high-level cognitive skills that allow us to plan, problem-solve, make decisions, and regulate our own actions and behaviors. The app uses scientifically validated tests to assess three different aspects of executive function: working memory; the ability to switch quickly between tasks; and inhibitory control, or the ability to override one's own habitual thoughts, behaviors, and impulses.

The Well Living Lab team deployed the app in studies of office lighting, and the results reveal how complex the environment's effect on cognition can be. Daylight and window views boosted employees' working memory and inhibitory control but had no effect on their task-switching ability. On the other hand, providing workers with artificial light that's on the cooler, bluer end of the spectrum—the type of light that typically tells our bodies that morning has arrived—improved their scores on the task-switching test, but not the other elements of executive function.

These nuances mean that a workspace that's just right for one enterprise or industry could be all wrong for another. When a job involves frequent disruptions and juggling many different kinds of tasks, as in an emergency room, it might be helpful to provide blue-enriched light. But in workplaces in which creativity is crucial, another lighting scheme could be optimal. In fact, a team of German researchers found that while cool light was ideal for performing tasks that required intense concentration, warm light was better for enhancing creativity. "There's no environment that's good for everything," Jamrozik told me.

Or for everyone. Despite what some titans of industry believe, employees aren't interchangeable cogs—we're individuals with diverse desires, sensitivities, and needs. An office that feels balmy to one worker can feel positively arctic to the colleague in the neighboring cubicle. In general, women tend to be more sensitive to temperature changes than men and to prefer warmer workspaces. An office calibrated to men's thermal preferences might make it hard for female employees to perform at their peak; women score best on cognitive tests at warm temperatures, while men do better at cooler ones, researchers found in a 2019 study.*

Similarly, an office that's designed to encourage spontaneous interactions and water cooler conversations might be energizing for extroverts but a nightmare for introverts like me. Introverts are more sensitive to noise and more susceptible to distraction than extroverts, scientists have found, and may have an exceptionally hard time working in open-plan layouts, which have become the dominant trend in office design.

Though employers have found a lot to like about open offices, which are flexible and cheap, employees almost universally hate them, complaining about the lack of privacy, the frequent distractions, and especially the noise. The relentless clacking of keyboards and murmur of half-heard conversations can make it difficult to stay focused and complete tricky cognitive tasks. (To add illness to injury, open layouts, where whole officefuls of employees are coughing and sneezing and shedding their microbes in close proximity, may make employees physically sick. In a 2011 study, Danish researchers found that people who worked in open offices took 62 percent more sick days than those who worked in private offices, which seem to provide the same protection from infectious disease that private hospital rooms do.)

Open offices may take a particularly big toll on certain kinds of workers—not just introverts, but also people with ADHD or autism,

---

*In the summer of 2018, when Cynthia Nixon was preparing to debate incumbent New York governor Andrew Cuomo in the Democratic gubernatorial primary, one of Nixon's aides requested that the thermostat in the debate hall be set to 76 degrees. In an e-mail to the debate organizers, the aide pointed out that work environments are "notoriously sexist when it comes to room temperature," *The New York Times* reported.

among others—and employers that insist on them would do well to provide some basic soundproofing, as well as at least a few places where overwhelmed employees can retreat and recharge.* Some companies are adopting activity-based offices, which allow employees to move between workspaces as their tasks change throughout the day. A staffer might use a standing desk to blast through e-mails, kick back in a comfortable lounge during a group brainstorming session, and then hole up in a private nook for focused research and writing. On the whole, surveys show that workers like activity-based offices and the autonomy that they enable. Whether these arrangements improve productivity and performance is less clear, but research at the Well Living Lab could reveal how to design spaces that support common workplace activities.

The Well Living Lab team has big ambitions, and no topic is off-limits. During the two days I spent at the lab, I watched the team constantly churn out new ideas. Perhaps they could turn the office space into a classroom, they mused, or study the health effects of shift work, or probe whether certain office conditions make it easier for people with traumatic brain injuries to return to work. They sketched out future studies that would investigate the health effects of office microbes, pollutants, and plants. "We're taking kind of a kid-in-a-candy-store approach," Bauer said.

When I checked in a year later, Bauer and his colleagues were beginning to unpack the best way to connect office workers with a healthy dose of nature. We've seen how powerful nature can be in helping to reduce pain in surgical patients or encourage school kids to be more active. Plants can also turbocharge mental performance, enhancing attention, focus, memory, learning, and productivity. Studies have shown that standardized test scores are higher at elementary schools surrounded by greenery and that students do better on tests of

---

*Steelcase, the office furniture company, collaborated with Susan Cain, the author of the bestselling book *Quiet: The Power of Introverts in a World That Can't Stop Talking*, to develop a collection of office spaces "designed for the unique needs of introverts." The prefabricated private workspaces provide visual privacy, soundproofing, and customizable lighting, among other features.

attention when their classrooms look out onto natural landscapes or feature plant-filled "green walls." (Offices have a lot of similarities to schools, and, in general, what's good for office workers—comfortable temperatures, minimal background noise, good ventilation, plentiful daylight, and views of nature—is good for students, too.)

According to Stephen and Rachel Kaplan, psychologists at the University of Michigan, nature's well-documented cognitive effects can be explained by what's known as the attention restoration theory. The influential theory holds that natural settings give the brain a break from the cognitively exhausting tasks—from memo writing to meal planning—that fill our daily lives. Nature draws our attention but engenders an effortless kind of engagement, often called soft fascination, that allows the mind to rest.

But is it enough to have a big nature mural on the wall and a tree or two in the office lobby? Or do we require a lush view from our desks and live plants throughout our workspace? Does adding a soft soundtrack of local nature sounds amplify the benefits? The Well Living Lab is trying to answer these questions.

Over the next few years, the team also wants to study whether dynamic, circadian lighting has benefits for office workers and plans to compare the effectiveness of several different sound-masking strategies. And, inspired by studies that have shown that office workers who feel more control over their surroundings report higher job satisfaction, they'll track what happens when they put their subjects in the driver's seat, allowing them to adjust the temperature, lighting, humidity, ventilation rate, and sound masking at their own workstations.

The facility itself will continue to evolve, and the lab's leaders have contemplated incorporating still more technologies, including facial and mood recognition software and pressure-sensing floor mats and furniture. They're opening a second Well Living Lab in Beijing, China. The facility is twenty-five thousand square feet—more than three times the size of the Minnesota lab—and capable of physically rotating to maintain contact with the sun.

The scientists hope to expand their reach with a suite of products that volunteers can set on their desks in their actual, real-world offices. That includes a small, low-cost, low-power sensor array capable of measuring air quality, light levels, noise, and more. Participants would wear biometric devices, perhaps smart watches or bracelets, and use their smartphones to complete short surveys and tests through the workday. All the data would be automatically uploaded to the cloud and sent back to the Well Living Lab. "What would be really helpful is to learn as much as we can in a highly controlled environment in our lab space and then take that learning and translate it into the *real* real world," Bauer said.

IT HAS TRADITIONALLY been difficult to do rigorous post-occupancy research, which has hampered our ability to understand what actually happens in offices and how they affect employees. "You can go to pretty much any company in the world and ask really basic questions about what goes on internally and they can't answer them," said Ben Waber, the CEO and cofounder of Humanyze, a Boston-based workplace analytics company. "Like, 'How much does management talk to the engineering team? How much do people work?' No one knows and that's crazy."

But technology is beginning to change that. Humanyze makes software and hardware that enables companies to analyze their employees' digital and in-person interactions. One of its products, known as the sociometric badge, can detect when two workers are having a face-to-face conversation. The badges, which are worn on a lanyard around the neck, are loaded with electronics. Each device contains a microphone, an accelerometer, and Bluetooth and infrared sensors, which can track the badge-wearer's location as well as the direction that he or she is facing. When two badge-wearers are in close proximity, facing each other and engaged in an alternating pattern of speaking, they're probably having a chat. (The badges don't record the content of conversations, and the accompanying software provides only anonymized, aggregated data, rather than information on individual employees, Humanyze says.)

Humanyze licenses its products to companies around the world, and Waber, who helped develop the badges as a graduate student at MIT, has used them to investigate what kinds of communication patterns lead to workplace success. In a study of IT employees, Waber and his colleagues confirmed that, despite the rapid adoption of messaging software like Slack, face-to-face communication remains the gold standard; real-life encounters are associated with higher productivity and performance, especially when the work is complex, and teams are more cohesive when they talk in person. They've also found that proximity is the best way to foster these interactions. "We see across companies that the likelihood you communicate with someone, face-to-face or digitally, is proportional to how close your desks are," Waber explained. It's yet another example of the fact that we are, at heart, creatures of convenience, most likely to eat food that's within reach and talk to people who are nearby.

Consider the case of a major European bank, which wanted to figure out why some of its branches were more successful than others. Using the Humanyze badges, the bank discovered that workers in its best branches were having more face-to-face conversations than those in its underperforming locations. What's more, in some of its low-performing branches, employees had sorted themselves into two distinct social groups, with members of each group rarely speaking to members of the other.

"Pretty quickly it became clear what was happening," Waber told me. In each of these branches, the workforce was split over two floors; employees on the first floor rarely went to talk to those on the second, and vice versa. "It takes less than ten seconds to walk up those stairs, but people don't do it," Waber said. "The organizing power of physical workplaces is so incredibly strong." In the subsequent months, the bank started systematically rotating its employees between floors.

The move paid off. Over the following year, sales at those branches increased 11 percent.

In another study, Humanyze found that an online travel company's highest-performing software engineers tended to eat lunch in big groups, while low performers dined with just a few colleagues. After the firm replaced its small lunch tables with larger ones, productivity jumped 10 percent.

Of course, fostering communication is complicated, and simply ensuring that employees are all sitting in close proximity doesn't necessarily mean they'll have meaningful conversations. That's what Ethan Bernstein, an associate professor of leadership and organizational behavior at Harvard Business School, learned when he used the Humanyze badges to help resolve an ongoing debate about the supposed benefits of open offices. Despite the well-documented drawbacks of open layouts, many executives continue to believe that demolishing the literal barriers in an office can help them overcome more figurative ones, that throwing all their employees together in a big open space can improve communication and teamwork.

Research on the subject, which has relied primarily on workers' own impressions of their office interactions, has turned up contradictory findings; the Humanyze badges gave Bernstein a way to measure worker communication directly. "If we're going to make claims about [how] open offices increase interaction, we should measure interaction, not perceived interaction—now that we live in a world in which we can do so," he told me.

Bernstein used the badges to track the employees of a Fortune 500 company that hoped to foster more in-person encounters by jettisoning its cubicles and moving to an open-plan layout. He recruited fifty-two of the company's employees to wear the sociometric badges before and after the redesign. Despite the company's intentions, in the new open office, face-to-face interactions plummeted; after the redesign, the amount of time that employees spent conversing in person fell by an astonishing 72 percent. However, digital communication, in the form of e-mail and instant messaging, shot up, suggesting that electronic interaction had replaced the live encounters that the company had been so eager to promote. (The company itself also reported that employee productivity fell after the redesign.)

There's plenty of psychological evidence that we desire at least a modicum of privacy in the workplace. In the absence of that privacy, employees might withdraw socially. Plus, they may be wary of having face-to-face conversations within earshot of their office mates. "And so very quickly people decide that the norm of interaction isn't really there, so what's the norm that replaces it?" Bernstein told me. "Put on your headphones, stare at your screen, work intently." And if you have a question for a colleague, just ping her online. "I don't blame anyone in that environment, quite frankly, for reducing their face-to-face interaction in favor of less risky electronic interaction," Bernstein said.

Bernstein's findings add an interesting wrinkle to the literature on open-plan offices, and the sociometric badges made it possible. "I love observational research, I love ethnographic research, I love qualitative research—all this is good work," he said. "But nothing really substitutes for exactly tracking the thing you're trying to study. I think that there are pluses and minuses to the technology, too, but by and large if we use it well, then this opens up huge opportunities."

ONE ORGANIZATION THAT made the most of these opportunities is WeWork, the controversial co-working company founded in 2010. WeWork rents out large office spaces, makes them over, carves them up, and then leases out desks and smaller workspaces to freelancers, entrepreneurs, start-ups, and even corporate behemoths like Microsoft and Facebook. A typical WeWork office includes a variety of spaces and amenities, including private phone booths and an assortment of meeting rooms, ranging from small lounges with comfortable seating to larger boardroom-like spaces with conference tables and whiteboards. Each location is unique, but they're all thoughtfully designed, with midcentury modern furniture, soft lighting, plentiful plants, custom murals, and playful graphic wallpaper. Some are dog friendly or have other perks like meditation rooms, fitness centers, outdoor terraces, and Ping-Pong tables.

WeWork grew at an exponential rate; by early 2019, there were more than 400 locations and 400,000 members globally, and its valuation had soared into the tens of billions. Along the way, it became a venture capital darling, but it also attracted intense scrutiny and criticism—for its unconventional, controversial founder; its reportedly frat-like culture; and, especially, the sustainability of its business model. WeWork was born in the aftermath of a global recession, when real estate was cheap, and its reliance on long-term leases could make the company especially vulnerable in the event of another downturn, critics say. And in the fall of 2019, WeWork's founder—who has been accused of unethical business practices, as well as unsavory personal behavior—stepped down as CEO, and the company abandoned its planned IPO in the face of deep investor doubts.

Though WeWork's future remains unclear, the bonkers growth and rapidly expanding portfolio that characterized the company's early years provided an opportunity to learn more about how people work and what kinds of office spaces they want and need. In September 2017, I went to WeWork's corporate headquarters, in Manhattan, to find out what the company had uncovered. The reception area was cavernous, with an eclectic collection of sofas and area rugs scattered across the hardwood floors. At 11:00 a.m. on a Wednesday, pop music was blaring, and dozens of hip young professionals were milling around. It felt more like a coffee house than an office, a vibe that was enhanced by the presence of an actual coffee bar, staffed by a barista. (Carafes of fruit-infused water were also available, and free craft beer was on tap.)

I barely had time to settle into one of the earth-toned couches before Daniel Davis, WeWork's slight and stylish director of research, appeared. After a quick tour, he ushered me into an intimate meeting room. We settled in at a small table, and Davis talked me through how WeWork transformed from a co-working company into a research enterprise.

As befits a modern company, WeWork relies on software to manage its spaces and interact with its members; over the course of a typical workday, tenants of every WeWork building generate trails of digital "exhaust" that contain clues about office behavior. "Architects often don't think of their buildings as these things that produce data or contain information about the world—they think of them as spaces and as experiences," said Davis, who has a PhD in computational design and the unhurried accent of his native New Zealand. But, he added, "The building is this thing that itself is producing data, and with that data, we have this ability to understand patterns and make predictions."

For instance, members book meeting rooms using the WeWork app; as of October 2018, the company's central database contained information on 6.8 million meeting room reservations. By studying this data, Davis said, "We're able to piece together a picture of how the spaces are being used."

The company has learned things that have changed the way they design—and even the way they conceive of spaces. "We used to describe meeting rooms as conference rooms, and I think even just that nomenclature suggests to a designer that it has to be this formal thing with a TV and a projector and, you know, there's going to be ten people looking at a PowerPoint presentation," Davis told me. But that turned out to be a relatively rare use case. The average meeting had just two to three people; even in rooms designed to accommodate twelve, 61 percent of meetings had four or fewer attendees. And in the vast majority of meetings, no one needed a projector or whiteboard.

The findings suggested that WeWork needed to provide more spaces for small groups to have casual get-togethers. "We still design spaces that accommodate those large groups having formal presentations, but we also started designing a whole series of spaces that are designed to enhance conversations, so the lighting's a little bit dimmer, and it's a little more intimate," Davis said. "The seating's generally less formal—so maybe a couch or some comfy chairs—and the acoustics in the space are different. There's no television or whiteboard. So that's something that empirically came out of that research and changed the way that we think about designing those spaces."

After a meeting has concluded, the WeWork app pings members, asking them to rate the room and leave comments explaining their ratings. "Someone will complain about a chair or they'll complain about the wallpaper or they'll say there's not enough whiteboard markers," Davis said. If the complaint is something that's easy to fix, as in the case of the missing markers, a WeWork community manager can remedy the problem on the spot.

Sometimes the comments reveal more fundamental design issues. In one Washington, D.C., location, the designers used what Davis describes as a "funky, bright yellow wallpaper" in one of the meeting rooms. "We see all this feedback for that room that's related to the wallpaper," he told me. "And all of it is incredibly negative—people saying that it's overwhelming and that it's distracting." Armed with that feedback, WeWork can switch out or paint over that wallpaper and make a note not to use it again. (The company keeps a running "blacklist" of design elements that haven't worked out, like a particular model of metal chairs that looked cool but turned out to be uncomfortable to actually sit in.)

Davis and his colleagues have also analyzed the leasing data on more than 3,000 private WeWork offices, spread across 140 buildings, to see if they could suss out why some spaces were in high demand while others had more trouble attracting tenants. Some of their findings were unsurprising—people preferred renting offices with windows—but others were less intuitive. They learned, for instance, that square offices were more popular than longer, more rectangular ones. "We think the reason for that is that a square room is easier to reconfigure," Davis told me.

As WeWork expanded around the globe, its designers and researchers began to explore cross-cultural differences in workplace behaviors. They found that the culture surrounding meetings varies by country—the gatherings tend to be larger and more formal in China than they are in Brazil—as does lunchtime behavior. American work-aholics often eat lunch alone at their desks, but in the Dutch WeWork spaces, it's common for everyone to sit down and break bread to-gether. This observation prompted the company to install bigger ta-bles in the kitchens and lounges in its Dutch locations. (Humanyze would approve.)

With these and other findings in hand, WeWork created software to help automate office design. The company developed an algorithm that predicts how often a given meeting room will be used, which helps its architects ensure that new offices have the right number and mix of meeting spaces. It also wrote a program that lays out desks in multi-person offices; the software, which maximizes the number of desks while simultaneously ensuring that workers have enough space to get around, can arrange twenty desks in less than a second. Both of these automated systems outperform human designers. "You could imagine in the future that these algorithms would be able to make more nuanced and informed recommendations about how the layout should happen in a space," Davis told me.

Whatever WeWork's future, I suspect its data-driven approach to workspace design will only become more common. "In ten or twenty years, there's going to be a lot more data about our built environment," Davis said. "It's inevitable that the fabric that makes up the built environment is going to be embedded with a lot more sensors and technology."*

But the more I learned about the power of this technology, the more I began to wonder who it's really for. Just because architects and employers have access to more information about office spaces doesn't mean that they'll mine it to workers' benefit. (A productive worker is not necessarily the same thing as a happy worker.) It's nice to imag-ine that companies would want to use an algorithm that helps them

---

*In fact, WeWork's approach is already spreading. Davis left the company in 2019 to create a new research team, similar to the one he led at WeWork, at a large, Australian-based architecture firm.

maximize the number of employees who sit near a window, but isn't it more likely that they'll use these tools to, say, figure out how to cram more people into a limited amount of space?

Beyond that, there's a fine line between research and surveillance. Many companies already use software to log their employees' keystrokes, monitor their arrival times, and chart their real-time locations, and new behavior-tracking technologies present new possibilities. Amazon has patented wristbands that track the hand movements of warehouse workers, and call centers have experimented with software that monitors the emotion in the voices of both their customers and their employees.

Companies could use these technologies to try to understand how to make their workers happy and what they need to succeed. Or they could use them to coerce and control their employees, and to wring every last ounce of productivity out of them. And I suspect that people who work in warehouses and call centers—and other settings in which laborers tend to be overworked and underpaid—are precisely those who are most likely to find themselves subjected to this kind of technological micromanaging. In addition to invading privacy, this surveillance can jeopardize worker health and safety as employees push themselves harder and harder to meet corporate-set metrics. Alternately, it can spur creative efforts to evade or fool the tracking technology—ultimately making it useless—or even backfire, making workers more stressed and less productive.

I asked Ben Waber at Humanyze about these concerns, and, to my surprise, he was quick to agree that workplace tracking technologies present serious risks. "Someone will eventually do something really wrong with this kind of data," he said. "There absolutely needs to be specific regulation around our industry." Digital privacy laws, like the European Union's General Data Protection Regulation (GDPR), are a good start, Waber said, but they're not set up to protect the rights of workers, who may find themselves under intense pressure to agree to technological monitoring. "For example, I can stop using Facebook,"

Waber told me. "It may be a little inconvenient, but you know what? Facebook doesn't control my life. It doesn't have much power over me. Your employer does."*

Ideally, Waber would like to see legislation that limits what kinds of data companies can collect on their employees and stipulates that when they test out new behavior-tracking technologies, they do it strictly on an opt-in basis. "There's a lot of benefit to this technology, but if it's done in a way that is micromanaging, that is Big Brother, you won't see all those good things," he said.

THERE'S ANOTHER WAY to increase the odds that this stream of workplace data benefits employees: put it in their hands. That's a possibility that Marc Syp began exploring a few years back, when he was leading the design computation team at the architecture firm NBBJ. (He has since moved on to a new firm.) Syp knew that the conditions inside an office could vary widely; the lounge near the kitchen might be warm, bright, and loud, while the alcove behind the meeting room was cool, dark, and quiet. But employees, he realized, didn't have a good way to monitor any of this. With just a few swipes on their phones, they could get a sophisticated readout of the weather outside; if they wanted to keep tabs on the indoor climate, they were mostly out of luck. "Buildings are really the last black boxes of the information age," Syp said. "You might as well drop your smartphone into a bucket of water as soon as you walk into a building."

Syp set out to remedy this problem by creating an app that permitted people to monitor the real-time environmental conditions inside their offices—and to find a place to work that felt just right. He called it Goldilocks. "The whole concept was, you're in the office, someone's talking on the phone, there's some glare from the window, and you

---

*In 2015, a California woman alleged that the wire transfer firm where she worked as a sales executive had fired her for deleting a cell phone app that tracked her location at all times, even when she wasn't on duty. She sued the company; the two parties ultimately settled out of court.

think, 'Okay, there's gotta be a better place to work,'" Syp explained. "You can fire up Goldilocks and say, 'I just need a place cooler and quieter.'" The app, which connects to environmental sensors scattered through the office, highlights the areas that currently meet those criteria, while motion sensors alert workers if a space they have their eye on is already occupied.

Some products are giving workers even more direct control over the office climate. Comfy, a California-based company, makes an HVAC-linked mobile app that allows office workers to tinker with the temperature of their own workspaces. An employee can tap the "cool my space" button and receive a blast of cool air; "warm my space" summons a balmy gust. The system learns workers' patterns over time: if a bunch of employees who sit by a big bay of windows request hot air when they clock in every day, Comfy will start automatically keeping that area warmer in the mornings. (The system saves energy, too. It turns out that setting the thermostat of an entire office to a single, predetermined temperature isn't just uncomfortable for workers—it's inefficient to boot.)*

Carlo Ratti, an architect and engineer who directs the Senseable City Lab at the Massachusetts Institute of Technology, has taken the idea of individual thermal environments even further. In the system he designed, heating and cooling machinery mounted to the ceiling creates personally tailored "thermal bubbles" that follow people as they move around a building.

These kinds of technologies will not "fix" the workplace. They don't address some of the major structural problems with modern employment, including overwork, stagnant wages, and the precariousness of freelance and contract work. Many of the world's laborers face

---

*Some engineers are pursuing other innovative strategies for personal temperature control. Scientists at the Center for the Built Environment at the University of California, Berkeley, have built and tested several "personal comfort systems," including electric, under-desk foot warmers and low-power office chairs that can provide heating or cooling to shivering or sweating workers. Researchers at the University of Maryland created RoCo, a mobile air conditioning robot that follows people around and bathes them in cold air.

problems that go far beyond a slightly chilly office or a too-formal meeting room, and even office workers would probably vastly prefer, say, better paid leave policies than perfectly programmed thermal bubbles.

Although WeWork itself explored the idea of super-personalized workspaces—during my visit to the company's headquarters, I saw a prototype of a smart desk that automatically rises to a member's preferred working height—Davis actually finds some of the potential applications to be gimmicks more than game-changers. "I question some of the value of that," he told me. "I'm not sure if this experience meeting with you right now would be significantly better if the temperature was five degrees different or the light was in the perfect shade of yellow that I wanted."

But the desire for this kind of personalization, and the endless innovation to make it possible, does reveal how far from ideal our current offices are. As technology becomes tightly integrated into our buildings, it could give at least some workers more control over their environments and empower them to create spaces that better meet their individual needs.

Of course, it's one thing to design an office that serves a diverse assortment of white-collar workers—it's another challenge entirely to create spaces that support the full spectrum of human ability and experience.

FIVE

FULL SPECTRUM

O N MAY 22, 2014, Lindsey Eaton took the stage at the Wells Fargo Arena in Tempe, Arizona. The petite blond high school senior loved writing and public speaking, and she had long dreamed of giving an address at her graduation. The moment had finally arrived. She stepped up to the microphone and delivered the speech she'd been practicing for months. "I have autism, which means I have a diagnosis of awesomeness," she told the crowd, to cheers and applause. "I want to thank each and every member of the faculty and every graduating student for seeing in me possibilities, not disabilities."

It was a triumphant moment, but the high faded fast. On the ride home from the ceremony, Eaton broke down in tears. Her classmates were all on their way to celebratory after-parties and then, in a few months, to college. She was headed home with her parents, with no clear plan for the future. All sorts of worries ran through her mind: How was she going to find a job? An apartment? Would she be able to live on her own?

In the months and years that followed, Eaton struggled. She watched her high school classmates and two younger sisters begin their lives as independent adults; meanwhile, she was stuck in her parents' home, worrying that her best days were behind her. This wasn't the life she had imagined for herself. "I had bigger dreams, I had bigger hopes, I had bigger expectations," she recalled when we spoke in the spring of 2018.

She wanted to find a job that she loved—at the time, she was hoping to become a preschool teacher—and to live on her own. But she didn't know how to make that happen. Eaton couldn't pay the rent on an apartment by herself, and she was nervous about the prospect of sharing space with a roommate. And there wasn't a place that would really meet her needs. She didn't drive, so she'd need to live close to public transit, but she couldn't tolerate loud noise, so a bustling downtown location wouldn't be ideal either. And because she was still trying to master some of the tasks of daily living, like cleaning and staying organized, she'd probably do best in a place with a robust support system or, at least, an understanding landlord.

Her parents reconciled themselves to never being empty nesters. "Our vision for Lindsey and her life was to have her live in a guest home behind our house," Doug Eaton, Lindsey's father, told me. "That was as expansive as our vision."

It's not an unusual story. Autism is a complex, heterogeneous condition; people on the spectrum are enormously varied in their skills, sensitivities, and strengths. (Hence the adage "If you've met one person with autism, you've met one person with autism.")* Some young adults with autism may require around-the-clock care, while others can join their neurotypical peers in college and sign the leases on their own apartments. Many find themselves stuck somewhere in between, wanting independence but struggling to achieve it.

Autistic adults are significantly less likely to live independently, and tend to be even more disconnected from their communities, than those with other kinds of disabilities, but that's slowly starting to change. And in the summer of 2018—four years after graduating from high school and a few months after we first spoke—Lindsey Eaton would finally move into a place of her own.

IN THE UNITED STATES, the disability rights movement took root in the mid-twentieth century, years before autism became a widely recognized condition. It gained steam in the 1960s as part of the broader social struggle to expand the rights of black Americans, women, and other marginalized groups. In the early years, discussions of accessibility focused primarily on people with physical disabilities. In 1961, the American Standards Association published a set of guidelines for "making buildings and facilities accessible to, and usable by, the physically handicapped," recommending design features like wheelchair ramps, wide doorways, and bathroom handrails. Many of these ideas were later formalized in the Americans with Disabilities Act (ADA), enacted in 1990, and the accompanying design standards. The ADA was a landmark piece of civil rights legislation, prohibiting disability-based

---

*The language around autism and identity is evolving. For years, it was considered best practice to use what is known as person-first language, describing someone on the spectrum as a "person with autism" to highlight the fact that their autism does not define them. And some people with autism, including Lindsey Eaton, continue to prefer this language. But many others say that autism *is* an important part of their identity and that they therefore prefer identity-first language like "autistic person." In light of these opposing views, I use "adults with autism" and "autistic adults" interchangeably, though I have respected all sources' preferences about how they are personally identified.

discrimination and mandating that buildings be accessible. (According to the act, a "failure to remove architectural barriers" in existing buildings was itself a form of discrimination.) It catalyzed major accessibility improvements, particularly for people who use wheelchairs: curb cuts, ramps, automatic doors, and accessible restrooms all became much more common.

But many barriers remain. Enforcement of the ADA is inconsistent, the law itself has loopholes, and many buildings and public spaces are still essentially unnavigable for people with disabilities. (Public transit, for example, is notoriously inaccessible.) What's more, designers, developers, and property owners have traditionally focused more on accommodating wheelchair users than people with less visible differences, especially those that primarily manifest themselves in the brain. Many features of the built environment can pose challenges for people with certain cognitive disabilities, mental illnesses, and neurological conditions. For instance, people with post-traumatic stress disorder (PTSD) can become anxious when they're forced to navigate narrow passageways or blind corners, while autism, epilepsy, migraines, and traumatic brain injuries can all make people exquisitely sensitive to certain sensory stimuli like light and sound. (That means that the open offices that employers seem to love so much can be a nightmare for people with these conditions, as can modern restaurants, which have grown unbearably loud.)

Beyond unwelcoming public spaces, people with some kinds of cognitive and developmental disabilities may have trouble finding homes that meet their needs, just like Lindsey Eaton did. "People end up in housing that doesn't really work well for them," said Sam Crane, the director of public policy at the Autistic Self Advocacy Network (ASAN). Flickering lights, overheard conversations, the buzz of home appliances, and the smell of a neighbor cooking dinner can all bother people with autism, who may struggle if they live in an apartment that's not soundproofed, for example, or that shares an HVAC system with other units, allowing external scents to seep in.

Moreover, some autistic adults may need living spaces that can accommodate their repetitive, self-soothing movements. Crane told me about a friend of hers: "She, like a lot of autistic people, has a need to jump up and down," Crane said. "So this is someone who really shouldn't be in an apartment that's above someone else." Her friend did manage to find an apartment that fit the bill but ended up with a landlord who hassled her for not keeping it clean enough. The friend ultimately moved out—and into Crane's basement. "Management practices can be really significant in terms of whether people can stay in an apartment or not," Crane said. Cost is a major obstacle, too, especially because many adults with developmental disabilities are underemployed and live on limited incomes.

In part, the lack of housing for autistic adults has been a chicken-and-egg problem. Relatively few autistic adults have lived independently because there hasn't been enough suitable housing or support, and designers have not traditionally prioritized their needs because there weren't more of them living independently. "There's quite a lot of people with physical disabilities who are living in the community and that created a pretty strong pressure to make housing physically accessible," Crane said. "It's been a little bit slower for people with really significant intellectual and developmental disabilities."

But thanks to several converging trends, ideas about accessibility are evolving. Over the last several decades, many countries have embarked on a process of deinstitutionalization, closing the large hospitals and institutions that were once filled with adults with mental illnesses and developmental disabilities. In 1999, the U.S. Supreme Court declared that sequestering disabled people in large group facilities was a form of discrimination and that government services should be provided "in the most integrated setting appropriate." As a result, far more adults with disabilities are living and receiving supportive services in their own homes, neighborhoods, and communities; many are advocating for themselves and fighting for their right to live, work, and attend school alongside people without disabilities. And researchers have accumulated evidence that when adults with disabilities live at least semi-independently, they have bigger and more diverse social networks, are more involved in their communities, report higher levels of personal well-being, and are more satisfied with their lives.

Additionally, disability rights activists have been advancing the neurodiversity paradigm, which holds that neurological conditions—including autism, dyslexia, Tourette's syndrome, and ADHD—are not defects or dysfunctions but simply different ways of experiencing the world, natural cognitive variations that come with some unique strengths. It's part of a broader cultural shift (albeit an incomplete one) in how we view disability. The traditional medical model of disability, which characterizes physical and cognitive impairments as problems to be fixed, has given way to the social model, which posits that it's not using a wheelchair or having autism that's disabling—it's living in an environment (and a society) that doesn't accommodate these kinds of differences.

"Accessible design" has given way to "universal design," in which the goal is to create spaces, products, and experiences that serve the widest possible range of people, at every age and along the entire spectrum of ability. The goal is to do more than simply grant people "access." Rather, it's to empower people to participate fully in all aspects of society.

"Everyone has a basic right to good design," said Magda Mostafa, an architect and associate professor at the American University in Cairo who specializes in designing for people with autism. "Design standards only cater to that perfect six-foot-tall male individual that has good visual capacity, has good hearing capacity, has a statistically typical sensory profile, and I think it's very, very limiting. We're excluding so many people."

Medical advances and our ever-increasing life spans mean that far more of us are living with disabilities than ever before. One in ten American adults reports having some kind of cognitive disability, and a number of cognitive and developmental conditions, especially autism and ADHD, are being diagnosed much more frequently than they were in decades past. And disability is dynamic; over the course of our lifetimes, we will all experience fluctuations in our physical and mental abilities.

Designers are increasingly taking these cognitive and sensory differences into account. For instance, when the architecture firm Perkins+Will recently designed a new building for the University of Cincinnati Gardner Neuroscience Institute, which treats people with an array of neurological conditions, it assembled a patient advisory group to learn how to make the building more welcoming. To aid patients who might have trouble with navigation, the architects made sure that the corridors provided views to the outside, which can help people stay oriented. To reduce glare, they wrapped the building in a white mesh that ensures that the daylight entering the building is soft and diffuse.

At Gallaudet University, which primarily serves students who are deaf or hard of hearing, architects and academics are pioneering a design approach called DeafSpace. The DeafSpace principles outline a number of design features that can make it easier for people to communicate visually, including translucent and partial walls, circular or semicircular furniture arrangements, and rooms that are painted in soft blues and greens, which contrast with human skin tones, making signs and gestures more visible. Wide hallways, slopes instead of stairs, and automated doors allow people to sign without interruption as they walk, while soundproofing can make spaces friendlier to people who use hearing aids or have cochlear implants.

And, around the world, there's growing interest in creating autism-friendly spaces, including schools designed specifically for autistic students and offices that meet the needs of a neurodiverse workforce. Zoos, aquariums, sports stadiums, amusement parks, movie theaters, supermarkets, and airports have created quiet zones and low-stimulation shopping hours and screenings. There's even a dedicated app, called KultureCity, to help people find these spaces and amenities.

("It's like Yelp for sensory inclusion," one of the app's creators told *Fast Company*.)

Designers and developers are exploring new housing ideas, too. "There's been increasing interest in understanding what adults on the spectrum need to live more independently in the community," said Sherry Ahrentzen, a professor at the Shimberg Center for Housing Studies at the University of Florida. A little more than a decade ago, Ahrentzen began looking into this subject herself, in collaboration with a nonprofit organization that wanted to start building housing for autistic adults. Fortunately for Lindsey Eaton, that organization happened to be based in Phoenix, Arizona, just miles from where she grew up.

IN 1991, DENISE RESNIK—a preternaturally cheerful marketing executive—gave birth to her second child, a boy she named Matt. Matt's infancy was unremarkable, but in the months following his first birthday, some of the skills that he'd mastered started to fall away. His language regressed, and he stopped making eye contact with his parents. By the time Matt turned two, doctors had diagnosed him with autism. "And we were told to love, accept, and plan to institutionalize him," Resnik told me.

Resnik, a Phoenix native who is the CEO and founder of her own communications firm, doesn't much like being told what to do, and in 1997, she cofounded what is now known as the Southwest Autism Research and Resource Center (SARRC). Over the years that followed, she helped SARRC grow into a 190-employee, $15 million organization that offers almost every imaginable service to autistic people and their families, including diagnostic testing, early intervention programs, educational workshops, support groups, a peer mentorship program, community outreach in English and Spanish, employment coaching, an inclusive preschool, and a research center.

As SARRC grew, Resnik kept thinking about housing, which was becoming an urgent issue: every year, roughly fifty thousand autistic kids come of age in the United States, and they all need places to live. The issue was personally pressing, too. Matt was growing up, and Resnik knew that she and her husband wouldn't be around forever. Though Matt loved to sing, he struggled with spoken language and experienced frequent seizures, which are not uncommon in people with autism. These challenges would make it difficult for Matt to live entirely on his own, Resnik thought, but she didn't think he belonged in an institution or a group home either. "I went to some of those built environments, and I ran away as fast as I possibly could," she recalled.

Resnik wanted something different for her son and for all the children, teens, and adults she'd gotten to know through SARRC. She began to consider creating a housing development specifically for adults on the spectrum, and in 2007, she partnered with Ahrentzen, then at Arizona State University, and her colleague Kim Steele. Together, they began to explore the options and lay out the possibilities, combing through the scientific literature and studying existing residences for adults with disabilities to learn more about what was and wasn't working. Ahrentzen and Steele used this research as a jumping-off point to develop a set of design goals and guidelines for creating homes for autistic adults. (They published their work in a 2009 report, *Advancing Full Spectrum Housing: Designing for Adults with Autism Spectrum Disorders*.)*

The guidelines are not hard-and-fast rules—there is no one-size-fits-all solution when designing for people with autism (or, for that matter, people without autism). For instance, while many autistic people are easily overwhelmed by sensory stimuli, others actually crave this kind of stimulation, which Steele knows from observing her daughter, who is autistic. "She really likes loud music, lots of movement in her body," Steele told me. "Whereas other people really have to have those

_____

*In collaboration with the Urban Land Institute, SARRC published a companion report, *Opening Doors: A Discussion of Residential Options for Adults Living with Autism and Related Disorders*, the same year.

noise-canceling earphones on at all times and really don't have that need to constantly be walking and bouncing."

Architects can easily tailor private homes to these individual preferences; building shared residences is trickier, but in most cases, designers should default to creating calming environments, Ahrentzen and Steele say. They can minimize sensory overload by using soft, muted color palettes, steering clear of loud patterns, and installing quiet appliances and HVAC systems.* They should avoid fluorescent lights, which can flicker and buzz, and may want to consider adding low-arousal "escape" spaces for people who are feeling overwhelmed. They can add dedicated sensory rooms, kitted out with colorful lights and tactile toys, for sensory-seeking residents, who can also embellish their own private spaces with the stimuli they need.

Ahrentzen and Steele's guidelines also urge architects to think carefully about residents' social lives. Because social interaction can be a challenge for some autistic people, there's often an assumption that people on the spectrum aren't interested in forming close relationships with others. In fact, Ahrentzen and Steele spoke to developers who didn't think they'd need to design any spaces for couples. But that's simply not true. "People on the spectrum have life experiences and want life experiences like a lot of other people not on the spectrum," Ahrentzen said.

To that end, the guidelines suggest that architects designing shared residences create common spaces—courtyards and kitchens, gardens and mailrooms—where residents can encounter one another. At the same time, they should find ways to make sure that people can control how much and what kind of interaction they have. For instance, designers could create common spaces with alcoves, nooks, and window seats, which make it possible for residents to spend time in the company of

---

*Calming doesn't have to mean drab or stark, Steele cautioned. "There's a tendency to err on the side of putting in these really blank rooms," she told me. "You get these stereotypes, the stereotype vision of what an autistic person is, and then people would design them a white box. That's not very helpful."

others without having to sit in the center of a crowd. They can strategically deploy half walls, wall cutouts, and interior windows that enable people to preview shared spaces before entering them.

Moreover, ensuring that residents have private spaces that they can call their own can make them feel more comfortable taking social risks. It also honors their dignity, something that group homes and institutions don't always do. "We saw places where they wouldn't put doors on bedrooms because they thought people could hurt themselves and always needed to be under surveillance," Ahrentzen said.

In addition, the guidelines suggest that housing for autistic adults should be durable and familiar, with homelike decor, easy-to-navigate layouts, and spaces that have clearly defined functions. It should promote independence and keep occupants healthy and safe, taking into account that some people on the spectrum have balance problems, visual impairments, or a tendency to wander. And it should be affordable and well integrated into the broader community.

The design guidelines were just a starting point for Resnik. In the years that followed, she and her colleagues at SARRC held a series of focus groups and discussions with autistic adults and their family members, autism service providers, local developers, and housing officials. In 2012, Resnik launched First Place, a sister nonprofit to SARRC; two years later, she closed the deal on a vacant 1.4-acre plot of land in the heart of midtown Phoenix. The organization set out to create an apartment building, which they called First Place–Phoenix, for adults with autism, and hired RSP Architects to lead the design process. The architects drew on Ahrentzen and Steele's report and the information SARRC gathered at its community meetings and focus groups; to solicit more feedback on their design ideas, they also hosted two national design charettes, both of which included autistic adults.

Along the way, Resnik decided she wanted to help give young adults the skills they'd need to thrive on their own, so she created the First Place Transition Academy, a two-year program designed to prepare autistic adults for independent living. Students in the program would live together at First Place–Phoenix and take classes at a nearby community college to help hone their life, social, and career skills.

Lindsey Eaton learned about the academy in 2016, shortly after its launch. She was euphoric when she was accepted, and that fall she moved out of her parents' house for the first time. (The First Place apartments weren't ready yet, so Eaton and her classmates lived in another local housing development, which they shared with about a dozen senior citizens.) It was a tough adjustment. Eaton found it hard to get used to a new schedule, and at first she felt anxious and alone, misunderstood by students and staff. But as the months wore on, her confidence grew. She made friends. She learned how to do laundry, go grocery shopping, and create a budget, and she challenged herself to try new things, like cooking and taking the local light rail. She learned how to advocate for herself and to ask her teachers for help when she needed it.

I spoke to Eaton the day after she earned her independent living certificate from the Transition Academy. The future she'd long dreamed of was finally beginning to come together, she told me. She'd landed a job she enjoyed, doing clerical work for the Arizona School Boards Association. She had a boyfriend who shared her love of football and Christian rock. And in just a few months, she'd become one of the first residents to move in to First Place–Phoenix.

WHILE RESNIK WAS working to bring First Place to life, researchers were trying to push the science forward, assembling more evidence about precisely how the built environment affects people with autism. One of those researchers was Shireen Kanakri, an assistant professor of interior design at Ball State University in Muncie, Indiana. In 2012, when Kanakri was in grad school, she spent time observing second- and third-graders at two schools for autistic children. She found that as classrooms got louder, the students began to display more signs of distress. They rocked, spun, and flapped their hands; repeated words over and over; hit themselves and others; and covered their ears.

When Kanakri was hired at Ball State, she wanted to build a lab that would allow her to run more controlled tests of how autistic children responded to sensory stimuli. It took her three years and $200,000, but in the fall of 2017, she launched her first study. On a drizzly November day, I flew to Indiana to see how it was going.

The lab, housed on the second floor of the university's applied technology building, looks like a pediatrician's waiting room, with the same motley assortment of adult and child-sized furniture. The decor is mostly subdued—taupe walls, nature-themed art, and earth-toned furnishings—but there are bright, playful touches: plastic bins full of toys, wooden racks stuffed with picture books; a big orange basketball pillow; a hopscotch rug.

The testing chamber, a small, soundproofed room, sits in the back corner of the lab. "It's a more controlled environment, so we can test whatever we want," Kanakri tells me. She can change the color of the chamber by pulling red, yellow, green, blue, or purple curtains across its walls. Both fluorescent and LED bulbs have been installed in the ceiling, as has a set of speakers; a camera, decibel meter, and light meter are mounted in one corner. Kanakri and her colleagues can watch everything happening in the chamber through a one-way mirror.

For her first study, she's gathering data on how different colors, lights, and sounds affect autistic children and their neurotypical peers. Each testing session takes about three hours, and on the day of my visit, two children are on the docket. Kanakri spends the morning observing a young neurotypical boy, part of the study's control group, as she changes the conditions inside the testing chamber. He doesn't appear to be bothered by noise, though he does exhibit some signs of stress when Kanakri switches from the LEDs to the fluorescent lights.

After a quick lunch break, Kanakri resets the lab for the afternoon test session. Minutes before it's due to begin, an autistic preteen boy in a blue fleece jacket comes darting into the lab, his parents a few steps behind him, and makes a beeline for the toy basket. He sifts through the toys and then paces around the lab, repeating a not-quite-audible phrase to himself. (Many autistic children engage in this kind of repetitive speech, which is known as echolalia.) His name is Henry; he is an eager giver of high fives and a lover of puzzles, McDonald's chicken nuggets, and YouTube.*

The first challenge is to get him wired up. Henry needs to strap a heart rate monitor across his chest and stick three adhesive electrodes to his bare skin. This equipment will track his heart rate, respiration rate, and blood pressure as Kanakri changes the conditions inside the chamber. Although the electrodes don't cause any physical pain, Henry—like many of the children she's tested, including those without autism—isn't thrilled about the idea of putting them on. "This is the hardest part," Kanakri says. Henry paces, plays with the zipper on his jacket, and murmurs to himself.

But his mother is a pro. She pulls out one of Kanakri's whiteboard easels, uncaps a dry erase marker, and makes a list of what they're going to do: "Stickers," she writes. She presses one of the electrodes to her chest, then tells Henry it's his turn. Henry tentatively takes an electrode and puts it just under his collarbone, as his mom has done. Then he puts on the other two. "Good job!" his mom says. "Good job!" Henry repeats.

His mom turns back to the whiteboard. "We're going to put a belt on," she says, and writes "Belt." She puts the heart rate monitor on herself and then says, "Okay, Henry's turn!" He puts the belt on without complaint. "Give me five," she says, and he obliges.

Henry and his mother head into the testing chamber, where Henry begins working on a puzzle. Kanakri closes the door and flicks off the rest of the lights in the lab. We sit at a table in front of the one-way mirror and look into the illuminated room. A large desktop monitor shows a live video feed of the chamber, while a neighboring laptop displays real-time data on the sound and light levels, as well as Henry's physiological data.

First up: sound. Kanakri starts piping forest sounds, mostly birds chirping, into the chamber. Henry repeats part of a nursery rhyme to

---

*The child's name and other potentially identifying details have been changed.

himself over and over again as he sorts through puzzle pieces, but he seems calm and focused. His heart rate is nice and low. "He's doing fine with that sound," Kanakri says. "He's happy."

The timer goes off, and the birdsong ends. After a five-minute break, it's time for a new soundtrack—the din of a busy restaurant. Henry isn't obviously distressed by the noise, but there's a noticeable spike in his heart rate. His repetitive speech becomes more pronounced. Kanakri is surprised by the noticeable deterioration. "It's not super noisy inside," she says.

The final set of sounds, loud highway noise, is even worse. Henry's heart rate ratchets up even higher. He picks up a bucket of blocks and tries to charge out of the testing room. His mother redirects him with a new puzzle, and he begins to calm down. "Good job?" he asks as he snaps two pieces together. "Good job," she reassures him.

For the next two hours, the researchers work their way through a variety of different conditions: blue curtains with LED lights, red curtains with fluorescent lights, and vice versa, and so on. Kanakri isn't sure what patterns will emerge at the end of the study, and she's prepared for the possibility that they won't be simple. Though it wouldn't be surprising if lots of kids are sensitive to noise and fluorescent lights, the prospect of finding some ideal wall color that will be good for all autistic children—or all neurotypical children, for that matter—seems like more of a stretch.

Even so, it's clear that people are eager for any empirical insights she can provide. There have been so many parents interested in enrolling their kids in the study that Kanakri expanded its size from seventy children to four hundred. She's been asked to help design several centers for autistic children based on what she's learning, and some parents have already repainted their children's rooms based on how they responded to the lab's colored curtains. "Parents want to do anything they can to make life easier for their children," Kanakri told me.

There's certainly precedent for the idea that the right design decision can make a difference for people with certain cognitive conditions or disabilities. There's now abundant evidence that a thoughtfully designed building can improve daily life for people with dementia, for example. One hallmark of Alzheimer's is a loss of navigational ability. "When they can't find their way, because the environment is confusing, they get anxious, they get upset, they get aggressive," said John Zeisel, the cofounder and president of Hearthstone Alzheimer Care, which runs residences for adults with dementia. "Those are not really symptoms of disease—they're symptoms of being in the wrong environment."

Senior homes can help these residents stay oriented by making sure that living units are relatively small and that the common areas, such as the kitchen, dining room, and activity room, remain highly visible as residents travel through the building. They can also ease wayfinding by minimizing hallway intersections, exit points, and changes in direction, as well as monotonous, repetitive design elements, like long corridors with dozens of identical doors.

Other studies suggest that simply being more thoughtful about how we design senior residences—and providing seniors with the same design features that keep us all healthy and happy, including plentiful daylight, soundproofed spaces, homelike interiors, and adequate privacy—can improve social interaction and engagement; reduce depression, agitation, anxiety, aggression, and psychosis; and even slow the cognitive decline of residents with dementia. "A large number of the so-called symptoms of Alzheimer's disease are a result of environmental factors that are not well suited to these people's needs," Zeisel said.

IN LATE APRIL 2018, with construction on First Place–Phoenix nearing completion, I fly to Arizona to join a "hard hat tour" of the building. Resnik arrives at the construction site in a sleeveless blue dress; she slips off her nude pumps and ties on a pair of black sneakers. When I stick out my hand to introduce myself, she pulls me in for an embrace. There are more than a dozen of us there for the tour, and we follow Resnik's lead, donning neon-yellow construction vests and protective headwear. She leads us into the lobby, which is still under construction. It smells like paint and plaster, and wires hang from the ceiling. "You are now standing inside the dream," she announces.

The dream is a four-story, 81,000-square-foot property with fifty-five apartments. The first floor will house staff offices and several four-bedroom suites for students in the First Place Transition Academy, while the upper floors contain one- and two-bedroom apartments, which can be leased for a year at a time. Prices start at $3,800 a month for a roughly 750-square-foot one-bedroom. The rent is steep, Resnik acknowledges, but includes all utilities and various supportive services: First Place support specialists are available 24/7 to help residents with whatever they happen to need, whether that's learning how to manage their medications or career and wellness coaching. (Some residents may qualify for additional government-funded services, like occupational therapy, meal delivery, transportation, or in-home aides who assist with bathing and dressing.)

We take a quick spin through the first floor, which is chock-full of the kinds of amenities that nearly any apartment dweller would covet. "You're not going to feel or hear anything that's institutional about this property," Resnik tells us. "This lives and breathes like any other property." There's an outdoor courtyard with a pool, a barbecue area, a community garden, and a large teaching kitchen, which will host regular cooking classes. Just beyond the lobby is a multipurpose community room that will host parties and events; a resident advisory council will take the lead. "Think about this space at night and on weekends," Resnik says, "and how that resident advisory council has now programmed the space for karaoke night, or maybe it's a talent show or maybe there's bingo or maybe there's a dance."

Then we head upstairs to see some apartments. As we clamber up the half-finished staircase, Resnik urges us to peer out the windows, which frame a view of the city. "Take a look out—but watch where you're walking—and recognize that these people who live here are part of a community," she said. "We are integrated very much into the fabric of this community." The building is just steps from a light rail line and a bus stop, and grocery stores, drugstores, museums, a theater, a library, a YMCA, and a bowling alley are all nearby. That's part of what appealed to her about this site; Resnik wanted residents to be surrounded by places where they could work, volunteer, and socialize, and she hopes they'll spend their days out in the community. In a few months, she tells us, First Place–Phoenix will have an open house, inviting local residents to stop in and get to know their new neighbors.

We gather at the top of the stairs on a large open landing. A darkened room sits to our right, behind a glass door. Each floor has a special activity room in the same spot, Resnik explains. This one will become a fitness room. The one just above us, on the third floor, will be a game room, stocked with both electronic and traditional games; the fourth floor will house a Zen meditation room.

Mike Duffy, the project's bowtie-and-workboot-wearing lead architect, chimes in. "One of the things that sets First Place apart is the dedication to common space," he tells us. In addition to the fitness, game, and meditation rooms, there are four lounge spaces on each floor, in an assortment of shapes and sizes; residents who don't feel comfortable in a crowd can play Scrabble with a friend in one of the smaller "pocket" lounges.

Resnik leads us down the hall and into one of the one-bedroom apartments. It has an open kitchen, with wood cabinets, a spacious pantry, and large white countertops. (The countertops are intentionally oversized so that residents can cook with friends.) The living room, which has enormous windows, is illuminated by the blazing Arizona sun. The bathroom is big and bright, with a fiberglass tub and sliding glass shower doors.

A lot of the most consequential design decisions are invisible. There's a 1.5-inch layer of gypsum concrete in between each floor, which will help dampen the sound of footfalls. The walls contain acoustic channels to further muffle sound. All the lightbulbs are LEDs.

There are some less conventional touches. In the shower, for instance, the plumber has installed the water and temperature control knobs on the wall opposite the showerhead, rather than directly underneath it. "It gives someone the opportunity to turn on the water without having to duck or reach through the stream," Duffy explains. "If they turn it on too hot, there is a little bit more of a buffer."

The entrance to each apartment is slightly recessed, set back a few feet from the hall, to allow residents to transition more gently between their private spaces and the building's more public ones. "So you're never going out the front door and into a major path of circulation," Duffy says.

And there are some technological bells and whistles designed to help keep residents safe. The stove and oven are connected to motion sensors and will shut off if they don't detect any movement in the kitchen for a prolonged period of time. Whenever that happens, an alert will go out to staff, who can work with any perennially forgetful residents on strategies for remembering to turn the oven off. (Residents can manually override the system if they're roasting a chicken, baking a pie, or making something else with a long cook time.)

The design team deliberately stayed away from more invasive monitoring technologies and chose not to integrate the latest smart home devices, like programmable lights and blinds, in part because they wanted residents to learn how to manage their spaces on their own. "We never wanted First Place to be the only place that someone could live," Resnik says.

The appeal of First Place–Phoenix is obvious. It's an attractive building, with top-of-the-line amenities, in the middle of a busy metropolitan area. It's the kind of place I'd love to live, and that's precisely the point. "One of my big takeaways is that design for someone on the autism spectrum is not that different from design for anyone else,"

Duffy says. "Good design for autism is very similar to just good design." Autistic people may be particularly sensitive to their environments, but dampening sound and avoiding flickering lights, balancing privacy and openness, and providing a medley of different kinds of spaces that occupants can choose from, depending on their specific needs—these are all design decisions that anyone can appreciate.

"One of the tenets of universal design is that if we can accommodate people with what we might consider 'extreme' differences, we can benefit the people who are considered 'typical' as well," said Madlen Simon, an architect and professor at the University of Maryland's School of Architecture, Planning and Preservation. "Autistic individuals highlight that for us because they can't tolerate some of the bad design that neurotypical people can."

Learning more about how people with disabilities perceive and respond to their environments—and how to create buildings that enable and empower them—will ultimately help us create spaces that are better for us all. The classic example is curb cuts, which were initially intended to help people using wheelchairs get onto and off of the sidewalk. But they've ended up making life easier for those pushing strollers or carts and riding bikes or scooters as well. The design goals and principles that First Place has deployed are similar, Resnik told me. "What we're looking at is neurological curb cuts," she said.*

First Place–Phoenix does have some limitations. The most obvious is cost; a $3,800-a-month apartment is simply out of reach for a lot of people. Students in the Transition Academy can apply for state funds to defray the cost of their tuition, which includes the apartment lease, and First Place has established scholarships to help students who are in need. "We're doing our best to help those families who otherwise

---

*In a 2015 study, researchers at the University of Southern California and Children's Hospital Los Angeles created a "sensory adapted" dental office designed to soothe children with autism; they dimmed the overhead lights, played relaxing music, and projected calming images onto the ceiling. These changes were soothing for both the autistic kids and their neurotypical peers, reducing anxiety, pain, and discomfort in both groups.

cannot afford to participate," Resnik said. But for now, the building's other residents have to pay the rent out of their own pockets. (Most get substantial financial support from their families, Resnik said.)

Residents must be capable of following basic rules, feeding and dressing themselves, and communicating in some form, and the development is unlikely to be a good fit for those who require extensive medical support or engage in violent or self-injurious behavior, Resnik said. Those criteria will invariably exclude some autistic adults, she acknowledged. But Resnik told me she spent years trying to figure out how to create a building that would serve everyone. Eventually, she realized that it wasn't possible. "That's how we landed on First Place," she said. "After spiraling for a while and trying to create something for everyone, we looked up and out, and made tough decisions on where we felt we could make our mark." And she's proud of what they've managed to accomplish. "We tried to create the biggest tent possible," she said.

In fact, there is no requirement that residents be on the spectrum. During the tour, Resnik tells us that she's hoping to attract people with a wide array of disabilities and that one of the building's first tenants will be a man with a traumatic brain injury. Several adults with Down syndrome have expressed interest, though none have signed up yet. "We are promoting and celebrating neurodiversity," Resnik says.

FIRST PLACE–PHOENIX IS part of a wave of new housing options for autistic adults. There's Sweetwater Spectrum, in Sonoma, California, which welcomed its first residents in 2013. Its 2.8-acre campus includes several shared homes and numerous amenities, including hot tubs and an organic farm. In 2016, the Dave Wright Apartments, an affordable housing development, opened in Heidelberg, Pennsylvania. Half of the building's forty-two units are occupied by adults with autism; the other half house low- and moderate-income adults who are not on the spectrum.* In Florida, the Arc Jacksonville Village, a thirty-two-acre gated community for adults with intellectual and developmental disabilities, held its ribbon cutting that same year.

Although these projects have had no trouble attracting tenants, these kinds of planned communities have also sparked controversy. The Autistic Self Advocacy Network, for one, opposes developments that primarily house people with disabilities. These projects, ASAN says, are a form of segregation, separating disabled people from the community rather than integrating them into it. Moreover, Crane told me, they're simply not a good use of resources. "Instead of using affordable housing dollars to build housing across the city that autistic adults can use, you've got this one place," she explained. If you'd prefer to live on the other side of town, you're out of luck.

In addition, Crane noted that many of the developments designed exclusively for people with disabilities impose undue restrictions on their residents and may outright exclude those who don't meet a set of narrow criteria. "A lot of planned communities will say that they're only open to people who are relatively independent," Crane told me. "And that's really concerning to us. You end up in situations where if you are autistic and you have significant support needs or if you have other disabilities like you use a wheelchair or you can't independently dress yourself and bathe yourself, that will then be used as an excuse to keep you out of the housing."

In light of these concerns, ASAN advocates for what is known as "scattered-site housing," in which autistic adults can choose from properties distributed across a city, living in the same buildings and neighborhoods that are home to people without disabilities. Increasing the overall supply of affordable housing and expanding access to housing vouchers for low-income people with disabilities would also be helpful, Crane said. "What a lot of people face as the main barrier ends up being just the stock of affordable housing," she told me. (As an added

---

*The developers of both properties drew on Ahrentzen and Steele's autism design guidelines, which the pair eventually turned into a book, *At Home with Autism.*

bonus, expanding affordable housing is good for everyone, whether or not they're on the spectrum.)

Crane also suggests that disability rights advocates engage with real estate brokers, landlords, public officials, and planners to ensure that more buildings meet the needs of autistic adults. "Urban planners are already thinking about the importance of affordable housing, but they're not necessarily making sure that the affordable housing has accessibility features," she said.

Finally, as the field of autism-friendly design grows, it is absolutely vital that designers include autistic people in the planning process. It sounds like a no-brainer, but far too often design teams will consult parents, caregivers, teachers, nurses, and therapists, but not autistic people themselves. "That means that a lot of the design features that they'll engineer in are more for the service providers than for us," Crane said. "You know, big open spaces that are easy to monitor, surfaces that are easy to clean." Consulting autistic people should be "priority number one," she added. "Get autistic people involved and all of these other things will sort of fall into place."

A development like First Place–Phoenix is not the be-all and end-all. Nor does Denise Resnik want it to be. Instead, she sees it as one step in expanding the portfolio of choices for autistic adults, one new option for a population that has far too few.

LINDSEY EATON HAD been so excited to move into First Place–Phoenix that her dad joked about camping out there overnight so that she could be the very first tenant. In the end, they just showed up on July 2, 2018, the day the building opened. She was one of about thirty residents who hauled their mattresses and stuffed animals and big plastic bins of belongings into the building that first week, joining their new neighbors in welcome barbecues and brunches and pool parties. Eaton selected a one-bedroom apartment on the second floor, with a 270-degree view of the city and easy access to a large lounge area. She put a photo of her colleagues at the School Boards Association on prominent display in her bedroom.

At first, Eaton and her parents weren't quite prepared for how different the First Place apartments would be from the Transition Academy. At the academy, she'd had a structured daily schedule and been closely supervised by staff. At the apartments, Eaton was far more self-sufficient. Support staff were readily available if she needed them—all she had to do was call or walk downstairs—but they wouldn't be checking in on her every day, making sure that her apartment was clean or that she was getting to work on time.

That was the whole point, of course, but the transition had been jarring. "It's what I dreamed of," she told me when we spoke that fall, several months after she'd moved in. "But it's got its kinks." She'd been surprised to discover that she missed having a roommate, as she did at the Transition Academy. She'd never lived alone before, and she wasn't sure that she liked it. "All of a sudden it's me in my one-bedroom apartment just talking to myself," she joked.

But she was adjusting to her newfound independence and starting to build a social life. She had joined the resident advisory council, which had been renamed the Council of Resident Engagement, and helped plan a Halloween party and a potluck Thanksgiving dinner. She had started using an app to track her expenses and was proud of how well she'd been able to stick to her budget; when she had a little extra spending money, she'd treat herself to a new wristband for her Apple Watch. "I feel optimistic," Eaton told me. Her parents agreed. "We feel like she could not be in a better place," Doug Eaton said. "We're so thankful every day that this exists."

Other residents were also finding their way. Before she moved into First Place–Phoenix, Lauren Heimerdinger, who has autism and is visually impaired, had been worried about striking out on her own. "I'm mostly nervous and scared to death about this," she told me a few months before the move. "I've lived with my parents for thirty-two years—that's all I know."

In the end, the transition had gone more smoothly than she imagined. She'd had to make a few modifications to her apartment, including adding braille stickers to the buttons on her microwave, but she was enjoying her independence, and the community that was forming in the building. "I'm a social butterfly, so I like to be around people," she told me. She'd started dating another resident and was leading Sunday-night meditation sessions in the building's Zen room. "I am proud that I've gotten this far," she said.

Still, Heimerdinger didn't want to stay at First Place–Phoenix forever; she hoped to spend a year or two there and then look for a more affordable place out in the wider community. "It's nice to have the staff here that will help you, but I don't want to have to check in with them," she said.*

When I called Resnik a few months after the apartments opened, she was pleased with how things were going. Her son, Matt, now in his twenties, was beginning to make his own transition to First Place–Phoenix. They were taking it slow. Matt, who had started a biscotti company to help raise money for his living expenses, was still working on some of the life skills he'd need to live on his own. (He didn't have the language skills required to attend Transition Academy, Resnik told me.) He'd started by spending two nights a week in his new apartment, a number that Resnik hoped would gradually increase. He was getting to know the property and learning the routine, doing yoga and playing Uno with his new neighbors. "It's exhilarating, it's joyful, and at the same time it's anxiety-producing," Resnik told me. But she and Matt were both beginning to conquer their fears, she said.

Resnik is working with researchers at Arizona State University to track how residents are faring: How do they rate their quality of life? Are they gaining life skills? How many of them work or volunteer in

---

*A few months after I first spoke with Heimerdinger, her father became the interim chief financial officer of First Place. It's not uncommon for parents and family members to join the organization. In the spring of 2018, as Lindsey Eaton was preparing to graduate from First Place Transition Academy, her father became a First Place board member. "I joined because I needed to see First Place succeed," he told me.

the community? Resnik hopes that if she can provide solid evidence that First Place–Phoenix is better for young adults than some of the alternatives, the government might provide public funding that will bring the rent down for residents. In the long run, she wants to create a portfolio of First Place properties; organizations and developers in several dozen cities across the United States and Canada have already expressed interest. "There's greater demand than ever before," Resnik said. "It would help us all to create more places and spaces where neuro-diversity can thrive."

Everyone deserves a safe, supportive place to live. And if we truly want to build a more equitable, inclusive society, that means extending the principles of good design not just to people of all abilities but also to people in all circumstances and settings. No matter what mistakes they've made.

JAILBREAKERS

I N RETROSPECT, PERHAPS they should have checked the fuel gauge before they stole the car. In 1993, when he was thirteen years old, Anthony Davis and a few of his friends swiped a gray Honda Civic off the streets of New York City. They weren't planning to keep it—they just wanted to go joyriding. And that's exactly what they did, speeding all over the Bronx. Until they ran out of gas. The police rolled up while they were trying to steal a replacement.

Davis had a turbulent childhood, one that he described as "a living nightmare." His father was absent and his mother was abusive, frequently burning Davis with a curling iron. When he was nine years old, he was sent to live with another relative, who turned out to be even more terrifyingly punitive. Davis and his two sisters ended up in foster care, where they were ultimately separated. Davis bounced around among foster homes and fell in with an older crowd, who introduced him to what he calls "the underworld" of drugs and crime. It wasn't long before he was caught stealing cars.

It was his first arrest, and it wouldn't be his last. Three years later, he was nabbed selling drugs and spent eight months on Rikers Island, which sits in the middle of the East River and houses a sprawling complex of New York City jails. His girlfriend was pregnant at the time; his first child—a girl—was born the day after his release. Six weeks later, his mother died from complications related to a blood disorder. "That changed me," he wrote in a letter to me. "Here it is a 16-year-old child goes from a boy to a motherless father in a matter of over a month."

He cycled in and out of jail until 2002, when he ran into a man from his old neighborhood. Years earlier, Davis told me, the man had sexually assaulted his older sister. "Just his presence bothered me," Davis said. "It's really tough to explain because I don't want to paint myself as a bad person. This is the lifestyle I was involved in." But one thing led to another. "We got into an argument, he said some things, I said some things, and I said, 'I'll be back,' and I left and came back, and I ended up shooting the guy." The gunshot wound was fatal. Davis took refuge at his girlfriend's house. When he returned home the next morning—planning to grab some clothes and flee the state—the police were already there. He pled guilty to manslaughter and was sentenced to twenty-two years in prison.

Shortly after beginning his prison term, he made a fateful mistake: he tried to break up a fight, jumping in between two men who were sparring. In the aftermath, all three inmates were punished, and Davis was sentenced to ninety days in solitary confinement.

Over the course of those three months, he spent twenty-two hours a day in a 105-square-foot cell. The concrete walls were painted a stark white. Everything else—the toilet, the sink, the bed, the desk, the shelves—was made of metal. The cell had a shower stall, which meant that unlike many men in solitary, Davis didn't even leave his cell to bathe. Or to eat; all his meals were delivered on a metal tray. Once a day, he was permitted to spend an hour alone in a small outdoor cage attached to the back of his cell.

Davis was young and brash, and he wasn't much troubled by being sent to what inmates often call "the box."* "Most people going into the box don't understand what they're in for," he told me recently. It was his first trip to solitary; at the time, he had no way of knowing that he'd spend seven of the next sixteen years in the box, or how profoundly it would change him.

I'd started writing to Davis because I wanted to understand more about how the built environment affects mental health, and prisons provide a frighteningly good illustration of how much damage the wrong environment can do. Most modern American prisons are intentionally harsh places. The buildings are literally designed to punish—to confine and control, shame and stigmatize, dominate and dehumanize. They are warehouses for human beings, isolating people from loved ones and throwing them together with strangers. They are stark and stressful environments—affording prisoners little privacy, freedom of movement, or control—and they are all-encompassing ones, to which people are literally captive.

It's no surprise that these conditions can take a serious toll on inmates, many of whom have histories of trauma, addiction, and mental illness. There are more people with serious mental illnesses in American jails and prisons than in psychiatric hospitals and hospital wards. Many come out in worse shape than when they went in.

A growing scientific understanding of the psychological damage that prisons can do has sparked calls for reform, and some architects are trying to create more humane correctional facilities, designing prisons, jails, and detention centers that are spaces of rehabilitation

*Solitary confinement is also sometimes called "the hole," administration segregation (AdSeg), or the SHU (pronounced "shoe," and short for Special Housing Unit or Security Housing Unit).

rather than retribution.* Of course, that's easier said than done, and the movement to create more humane prisons reveals as much about the limitations of evidence-based design as it does about its promise. Redesigning our prisons won't solve some of the most fundamental problems with the criminal justice system—which locks away far too many people for far too long—but it is an opportunity for us to rethink how we treat the people who are caught up in it. Architecture gives us a chance to express our values, to decide what kind of society we want to be, and the idea of "humane" design applies far beyond prison walls. Designers of prisons, hospitals, schools, and even entire cities are increasingly wrestling with the same big question: What does it mean to design for human dignity?

ALTHOUGH IMPRISONMENT IS a bona fide industry in America to-day, with more than 2 million adults behind bars, it's a relatively new form of punishment.† For most of history, accused criminals were de-tained only until their actual sentences—beatings, banishment, hard labor, or execution—could be meted out. Accordingly, imprisonment was ad hoc, with accused criminals stashed wherever there was space. (In medieval Europe, this was often in castle dungeons.)

Dedicated detention facilities proliferated during the sixteenth through eighteenth centuries. Many of these jails and prisons were just large holding pens that threw everyone together in large communal cells. Violence was common, and the welfare of prisoners was of such little concern that some died of starvation. The top priority was simply to keep inmates from escaping. Designers met this imperative primar-ily through the use of strong physical barriers: thick stone walls, heavy doors, strong metal bars. The spaces were unsanitary, with little access

---

*In general, jails hold people who are awaiting trial, while prisons house those who have been convicted and sentenced. When I use phrases like "humane prison design," I'm referring to efforts to reform and redesign all kinds of correctional facilities, not just prisons.
†Globally, there are more than 10 million people being held in jails and prisons, according to a 2015 report. The figure represents a 10 percent increase since 2004.

to daylight or fresh air, and deliberately imposing. As one eighteenth-century architect explained, a correctional facility should be "short and massive, where the prisoners, humiliated, weighted down, are constantly before the eyes of other criminals who are confined there, offering a vision of the punishments that await them, in the repentance that must follow the dissoluteness of their past life."

But as the eighteenth century wore on, reformers, who were appalled by the conditions inside prisons and viewed corporal and capital punishment as inhumane, began to agitate for change. They argued that detention itself could be used as a form of punishment, and that criminals could be rehabilitated during their incarceration. In Pennsylvania, the Quakers came to believe that inmates could be reformed through extreme isolation. Solitude, they argued, would prompt inmates to reflect, repent, and grow. And, at Eastern State Penitentiary, which opened in Philadelphia in 1829, they decided to enforce this isolation through architecture.*

In some ways, the inmates' quarters could be considered first-class; each man had his own private cell—complete with running water and a flush toilet, innovations that hadn't yet made it to the White House—that opened onto an individual, enclosed outdoor rec yard. But these features only served to reinforce the men's isolation. They never visited a communal toilet or cafeteria, never mingled with other prisoners outdoors. On the rare occasions they left their cells, they wore hoods to prevent them from even glimpsing their fellow prisoners. Silence was strictly enforced; the inmates were forbidden from singing or whistling, and the guards wore socks over their shoes to quiet their footfalls as they made their rounds.

Unlike many other penitentiaries, Eastern State was peaceful and orderly, and what came to be called the "Pennsylvania model" was quickly adopted in Western Europe. But it soon became apparent that extreme isolation was not therapeutic. At these facilities, doctors and

---

*The prison, which still stands, remained in operation until 1971 and is now a popular tourist attraction.

prison inspectors reported, inmates cried, trembled, and hallucinated; they became irritable, agitated, and manic. In his 1847 report, *Prison Discipline in America*, Francis C. Gray wrote, "It appears that the system of constant separation as established here, even when administered with the utmost humanity, produces so many cases of insanity and of death as to indicate most clearly, that its general tendency is to enfeeble the body and the mind." By the late nineteenth century, solitary confinement had largely fallen out of favor.

Although the Pennsylvania model had been born here in the United States, it never became quite as popular at home as it was abroad. Instead, while European countries were busy trying to replicate Eastern State, many American localities were implementing a rival model, which had been developed at Auburn State Prison in New York. Officials at Auburn, which opened in the early nineteenth century, came to believe that the best way to reform prisoners was not through isolation but through strict discipline and manual labor. Though Auburn inmates lived in single cells, they ate their meals in a shared dining room and spent their days together in the prison's workshops, where they made furniture, clothes, shoes, buttons, nails, barrels, combs, brooms, buckets, and other products. (For a brief period in the mid-nineteenth century, prison officials even brought in mulberry trees and silkworm cocoons and had the inmates spin silk.) The rules at Auburn were draconian; the men were given buzz cuts, dressed in matching black-and-white-striped uniforms, and made to march in lockstep, with their heads bowed, when they moved throughout the building.

Officials at Auburn argued that inmates fared better under this system than under strict isolation, but the model's real draw was financial; if architects didn't have to give every inmate his own spacious cell and private yard, they could build taller, more compact prisons, with cell blocks stacked atop one another. Plus, states could profit off inmates' labor by selling the goods that they produced. These economic advantages made the Auburn system irresistible to many government officials.

As the national prison population grew, however, the limits of the approach became clear. Prisons started housing multiple people in each cell, making it harder to maintain strict order. For all reformers' efforts, detention facilities became chaotic, violent, and overcrowded, just as early jails had been. And in the late twentieth century, solitary confinement would make a dramatic comeback.

ON THE MORNING OF October 22, 1983, three guards at the United States Penitentiary in Marion, Illinois, were escorting Thomas Silverstein back from the shower. Suddenly, Silverstein stopped and reached into another inmate's cell. With a few quick motions, the prisoner helped Silverstein slip out of his handcuffs and presented him with a homemade shank. Silverstein attacked one of the guards, stabbing him to death. Hours later, inmate Clayton Fountain managed to pull off the same maneuver—enlisting a confederate to pass him a shank and help him break free of his cuffs—before fatally stabbing another guard.

At the time, the Marion penitentiary was the designated home for the nation's most dangerous criminals. Silverstein and Fountain—and other men who were considered especially high-risk—were confined to the prison's "control unit," permitted to leave their small, single cells only when shackled and escorted by multiple guards. It was supposed to be the most secure unit at the most secure prison in the United States. And yet, over the course of a single day, two guards had been brutally murdered.

After a prisoner was killed a few days later, Marion put all its inmates, even those who hadn't been assigned to the high-security control unit, on lockdown, confining them to their individual cells for twenty-three hours a day. Family visits were sharply curtailed, the law library was closed, and religious services were suspended; the prison chaplain gave Communion to prisoners through their cell bars.

But the lockdown wasn't lifted after the prison was secured. Instead, it stretched on for weeks, and then months, and then years. What started as a temporary state of emergency became a permanent way of life, formalized through a series of design changes. The prison administration replaced inmates' beds with concrete slabs. It renovated the visiting room, installing Plexiglas booths that prevented prisoners from coming into physical contact with their loved ones. It bulldozed the fish pond and gardens, and divided the large rec yard into smaller, caged-in spaces.

The new regime, which ultimately lasted for twenty-three years, was harsh and dehumanizing, but it did impose order. The rate of inmate violence dropped, and when a prison reform group visited Marion in the fall of 1985, its inspectors acknowledged that "an uneasy calm prevails." For many correctional officials, Marion became a model, a way to bring America's chaotic prisons under control.

When the Marion lockdown began, the American correctional system was in crisis. The prison population was exploding, prison gangs were gaining power, and assaults—on both inmates and staff—were on the rise. In the face of all this turmoil, the system of strict isolation and control that emerged at Marion seemed like just what the doctor ordered. The supermax building boom was on; throughout the late 1980s and 1990s, dozens of super-maximum-security prisons, which confined inmates to their individual cells for nearly twenty-four hours a day, sprang up across the United States. Even low-security facilities added super-secure isolation cells and units—sometimes called "prisons within prisons." Solitary confinement became a widespread feature of the American penal system, and the number of prisoners held in some form of solitary exploded.*

Although there is no official tally of the number of prisoners in solitary confinement, experts estimate that the American correctional

---

*Although the United States has the dishonorable distinction of being a global leader in solitary confinement, the practice is used all over the world, from Ireland to Iran. In Scandinavia, which has often been praised for its humane prisons, detainees are routinely placed in solitary while they're awaiting trial. And in 2016, *The Guardian* reported that a number of European nations were increasingly using solitary confinement to hold prisoners who were accused or convicted of terrorism-related acts.

system has as many as a hundred thousand people in solitary at any given time. Some inmates are held there for their own protection—because they are at high risk of being assaulted by other prisoners, for instance—while others are sentenced to temporary terms in solitary for breaking prison rules. Inmates who are considered especially dangerous, difficult to control, or at high risk of escape may spend years, even decades, living in solitary. (After murdering the Marion prison guard, Silverstein spent nearly the entirety of the next thirty-six years in solitary, until his death in 2019.)

For Anthony Davis, the man who first found himself in solitary after trying to break up a fight, life in the box was "shell-shocking," he told me. It made him angry and defiant, and over the years that followed, he kept getting into fights with guards—and landing in segregation cells in prisons all over New York. In many of his letters and phone calls to me, Davis was warm and engaging; he told me about saving an injured bird in the prison yard, shared at-home cold remedies when he heard me sniffle, and called me excitedly when he learned that he had just become a grandfather.

When our conversations turned to solitary confinement, however, he became mournful. Prison is a place of tough guys and big egos, he told me, but he wasn't too proud to admit what solitary had done to him. Each time he went back to the box, it broke him down a little more. In solitary, he was allowed few personal possessions—usually just a handful of letters, pictures, and books—and he tried to pass the time by reading and writing. (During one trip to solitary, he wrote a novel, the first of what he imagines as a trilogy. "It's about three women in New York City," he told me. "Like a *Sex in the City* type of thing.") But the boredom and the loneliness were crushing. "Sometimes my mind was just as stagnant as my body," he wrote in one letter in his small, tidy handwriting.

As time passed, his thoughts became darker. "When your life is subjected to a small room, your brain kind of races and races," Davis explained. He became delusional and paranoid. He convinced himself that everyone, from the guards to his loved ones, was out to get him. He'd blow up over small slights. He became aggressive and confrontational, screaming, cursing, crying, and punching the walls so hard his hands would bruise. He felt like he was at war with himself. "I felt like a caged animal," he wrote. "It was like my insanity and rage wanted [to] burst out of my skin every second of the day."

Davis's experience is typical. When Craig Haney, a psychologist who studies the effects of incarceration and isolation, interviewed one hundred randomly selected men in solitary confinement at California's Pelican Bay State Prison, he found that 88 percent reported experiencing "irrational anger." More than 80 percent reported anxiety, intrusive thoughts, oversensitivity to stimuli, confusion, and social withdrawal, while roughly three-quarters reported mood swings, depression, and a general "emotional flatness." Seventy percent "felt themselves on the verge of an emotional breakdown." The men also reported an array of physical symptoms—including headaches, heart palpitations, sweaty hands, insomnia, nightmares, loss of appetite—that often accompany psychological stress and trauma. Significant proportions of prisoners displayed signs of more serious psychopathology; more than 40 percent experienced hallucinations or perceptual distortions. "There are few if any forms of imprisonment that appear to produce so much psychological trauma and in which so many symptoms of psychopathology are manifested," wrote Haney, a professor of psychology at the University of California, Santa Cruz.

It's difficult to prove that solitary confinement *causes* all these symptoms, and research is complicated by the fact that inmates who are mentally ill are sent to solitary confinement at disproportionately high rates. But there's plenty of incriminating circumstantial evidence: the symptoms occur even in prisoners who are healthy when first sent to solitary, they tend to get worse the longer people spend there, and they often ease when solitary confinement ends. "The conditions themselves are just toxic to human beings," said Terry Kupers, a California-based psychiatrist who has served as an expert witness in class-action lawsuits over conditions in American prisons. In a 2014 study of New York City jails, researchers found that inmates who spent time in solitary confinement were nearly seven times more likely to harm themselves than those who never went to the box, even after controlling for serious mental illness.

For Davis, the bottom came in the summer of 2013. He'd recently been sentenced to three years in solitary—the term would later be reduced to two years—after getting into yet another fight with a guard. He'd been moved to a supermax prison hours away from his family, and his loneliness overwhelmed him. Something about this specific cell felt especially oppressive. "The walls were really beige, and it was just—I felt dead," Davis told me. "I didn't feel like I was living."

One morning in early August, he decided that he'd had enough. "I just woke up feeling some kind of way," he recalls. When the guards escorted him to the shower, he asked for a razor so he could shave. He sat down on the floor of the shower and slashed his wrist. The prison rushed him to the hospital; he refused anesthetic when the doctor stitched up his wound.

And then he went back to solitary to finish his sentence. "There was a rapper and he said, 'They tryna kill me at the same time keep me alive,'" Davis told me. "That's a perfect example of what the box makes you feel like—it makes you feel like you're being killed and kept alive at the same time."

WHAT IS IT, exactly, about this kind of confinement that causes human beings to deteriorate so dramatically? One possibility is that solitary does its damage by depriving people of much-needed sensory stimulation. In the 1950s, researchers at McGill University in Montreal demonstrated the harm that this kind of sensory deprivation can cause, paying college students $20 a day to lie on a bed in a small cubicle. The students wore frosted eye masks, cotton gloves, and cardboard sleeves on their arms. A U-shaped pillow covered their ears, and the hum of an air conditioner provided constant white noise.

After just a few days in the cubicle, the students became irritable and restless, their thoughts muddled. Like many people in solitary, they developed delusions—some felt as though the researchers were out to get them—and experienced increasingly elaborate hallucinations, including "prehistoric animals walking about in a jungle" and "processions of eyeglasses marching down a street." Some students heard voices; others felt phantom sensations. One reported, "Something seemed to be sucking my mind out through my eyes."

Prisoners in solitary don't experience such extreme sensory deprivation, but they are subject to an unrelenting sensory sameness. One inmate I corresponded with, Francis Harris, has been in solitary continuously since 1997. (Harris was convicted of first-degree murder and sentenced to death in Pennsylvania; the state has historically kept all its death row inmates in solitary confinement.)* That means that for the last twenty-plus years, he's spent almost every hour of every day staring at the light gray walls of a cell that's smaller than the average parking spot. The window at the back of his cell looks out onto the prison's razor-wire fencing. Another small window in the cell door provides a view of a brick wall. "I miss nature," he told me. "I have not stepped foot on grass or dirt in two decades."

This kind of sensory monotony can, unsurprisingly, cause stress. As Colin Ellard, a neuroscientist at Canada's University of Waterloo, explained, "The best way to think of how people engage with environments is to remember that we're sort of info-vores, that we thrive on new information. We crave variety."

Ellard has demonstrated this preference in his own research, which involves leading volunteers on guided walks through the streets of major

---

*In November 2019, Pennsylvania agreed to end this practice as part of a settlement of a class-action lawsuit; as of January 2020, the change had not yet been implemented, Harris told me. Harris has also sued the state's department of corrections for failing to provide him with medical treatment.

cities, including New York City, Toronto, Vancouver, Berlin, and Mumbai. During these walks, the participants recorded their impressions and wore biometric sensors that tracked their eye movements, recorded their brain waves, and measured their skin conductance, which is a rough proxy for physiological arousal. When they stood in front of building facades that were low in visual complexity—imagine, for instance, a "big box" store with a blank, featureless exterior that takes up an entire block—their arousal levels dropped. They also reported being less interested, and taking less pleasure, in their environments. In other words, they seemed to get bored. "That's a little bit concerning," Ellard told me, because "boredom can actually produce stress." (The neuroscientist James Danckert, one of Ellard's colleagues, has shown that watching even a short boring video can trigger an increase in the stress hormone cortisol.)

In addition to lacking positive stimulation, the sensory experiences that inmates *do* get in solitary are often extremely unpleasant. Harris's first cell was overrun with mice and roaches, he said. "The mice had an attitude," he wrote to me. "Usually, if you turn on the lights you expect the mice and the roaches to run for cover, but these guys would just look over their shoulders at you as if to say 'Do you mind? I am trying to eat your food here.'" And the smell in that unit, he told me, was overpowering, a mixture of human excrement and body odor. "It is so strong that you gag when you first get there," he wrote. His current cell is much cleaner, he said, but he still has to contend with loud noise—the sound of other men screaming, the thunder of the guards' boots on the metal walkway above his cell—at all hours of day and night.

But as unpleasant as these sensory stresses can be, it's likely the extreme isolation of solitary that does the most damage. Humans are a social species, innately predisposed to connect with others. "We require social relationships as just part of being human," Kupers told me. "To take it away is just cynical and callous."

Loneliness and social isolation are associated with elevated levels of stress hormones, high blood pressure, inflammation, altered gene expression, and poor immune functioning. People who are—or feel—

isolated are at increased risk for heart disease, age-related cognitive decline, an array of mental disorders, and early death. Though it's possible that poor health causes isolation, rather than the other way around, animal experiments provide compelling evidence that isolation itself can in fact cause neurological and behavioral changes. Animals subjected to conditions that resemble solitary confinement develop behavioral abnormalities that mirror what we see in humans, including depression, aggression, and self-mutilation.

Even the prisoners who survive their time in the hole may be left with lasting scars. After multiple bouts in solitary, Davis now finds it difficult to be around other people. He still has a hair-trigger temper and erupts over minor provocations. His rage has hurt his relationships with friends and loved ones, and he doesn't always recognize—or like—the person he has become. "The charming, funny, handsome, charismatic individual whom I used to be is gone," he wrote in a 2014 essay. "His soul snatched away by the psychological effects of solitary confinement. I have now become this soul-less, bitter, fraction of myself who is emotionally unstable and full of rage. This is not how I want to live; who I want to be."

In 2011, a United Nations official declared that solitary confinement "can amount to torture," and a broad global consensus is emerging that solitary should be used only as a last resort and then only for short periods of time. (Experts generally recommend 15 days as the absolute maximum and say that children and people with mental illnesses should never be subjected to solitary.) Although the U.S. Supreme Court has so far declined to outright ban the practice, some states have imposed new restrictions on its use and a few have even closed down segregation units and maximum-security facilities. (Because high-security prisons are expensive to run, decommissioning them can save states substantial amounts of money.) And in 2018, Congress strictly curtailed the use of juvenile solitary confinement.

Although solitary confinement is a uniquely dangerous practice, efforts to rein it in are part of a larger movement to reform the American criminal justice system. Incarceration rates in the United States skyrocketed between the 1970s and the 2000s. (They have finally started to decline, albeit modestly, over the last decade.) The United States comprises less than 5 percent of the global population but accounts for nearly 20 percent of the world's prisoners. Many people receive harsh sentences for relatively trivial crimes, such as marijuana possession or parole violations. Others spend weeks, months, or even years behind bars simply because they can't afford bail. People of color are imprisoned at disproportionately high rates, and the percentage of inmates with mental illnesses has risen dramatically over the past few decades. These troubling facts have given rise to a rare bipartisan call for sweeping change, including revising sentencing guidelines, eliminating cash bail, and developing alternatives to incarceration.

Some designers are also rethinking prisons themselves, which can be isolating, dehumanizing, and unsafe, even for people who never set foot in solitary. American prison reformers have been inspired by some of the "humane" correctional facilities that have emerged abroad. The archetypical example is Halden Prison, nestled in the pine woods of southeastern Norway. Although a twenty-foot-high wall rings the maximum-security facility, the complex is designed to look and function like a small village. There are no bars on the windows, and the inmates are free to wander among the prison's buildings and common areas. There are multiple gardens and dozens of art installations, as well as a tree-lined walking trail, a rock-climbing wall, a ceramics workshop, and even a recording studio. The inmates have bright private rooms, complete with en suite bathrooms, miniature fridges, and TVs. The goal isn't to coddle the men but to nurture and rehabilitate them, preparing them for healthy, productive lives in society upon their release.

Halden has attracted considerable praise and attention, as have other similar facilities throughout Scandinavia. But these societies— and their criminal justice systems—do not much resemble America. Norway, for instance, has no death penalty and essentially no life sentences. The country is also relatively egalitarian, with a robust social safety net and a comparatively tiny prison population. Could a place like Halden ever work in the United States? As I began to research that possibility, one name kept popping up: Las Colinas.

THOUGH IT STANDS alone on a dead-end road, the Las Colinas Detention and Reentry Facility, in Santee, California, sneaks up on me. There's no looming wall or imposing guard tower. One minute I'm driving down a quiet, tree-lined street, with desert hills rising in the distance, and the next I'm at the end of the road, having entirely missed the turnoff.

After a quick U-turn, I pull into the parking lot. I step out of the car and start shedding my valuables—I'd been asked to leave my cell phone and wallet behind—when a car pulls into the spot next to me. Out steps James Krueger, a design principal at HMC Architects. He is a vision in gray: light gray shirt, dark gray jeans, hair shot through with a smattering of silver. It's a gorgeous Southern California day in the spring of 2018, and Krueger, who led a major redesign of Las Colinas a few years ago, is here to show me the utopian future of American jails.

The original Las Colinas opened as a juvenile detention facility in 1967 and became a women's jail twelve years later. By the late 1990s, it was in need of a major makeover. "It was very old, it was very dark, it was very gray, there was a lot of fencing, a lot of what they call razor wire," said Christine Brown-Taylor, the reentry services manager for the San Diego County sheriff's department, which runs Las Colinas.

In 1999, San Diego County hired Carter Goble Lee (CGL), a correctional consulting and planning firm, to assess the county's correctional needs and draw up a master plan for modernizing its facilities. Renovating Las Colinas emerged as the top priority. Together, CGL and the sheriff's department decided that they wanted to design a women's jail that felt less institutional. "From the very outset, we had a staff from the sheriff's office that were committed to doing something different," Stephen Carter, the executive vice president of CGL, told me.

The county issued a call for proposals, asking architects to submit designs for a secure complex that would allow detainees to socialize, congregate, and move around with few physical barriers. It caught the attention of HMC Architects, a local firm that thought it could leverage its experience designing schools to create a complex that felt more like a college than a jail. HMC partnered with KMD Architects, a San Francisco–based firm with expertise in designing correctional facilities.*

The HMC-KMD team won the contract with a concept for an open, parklike campus with meandering walking paths and numerous outdoor recreational spaces. They put the amenities—including a large outdoor amphitheater and grassy areas that they imagined being used for Frisbee and soccer—front and center and made the security features deliberately inconspicuous, hiding the perimeter fence behind careful landscaping. They added stand-alone buildings for classes, vocational training, and religious worship, hoping to create a place where women could "hit the reset button and really reboot their lives," Krueger explained.

At least, that was the hope. Krueger hadn't been back to see Las Colinas since it opened in 2014, and before we started our tour, he admitted to feeling a few jitters. "I'm a little worried we're going to walk in today and it's going to feel like a prison," he told me. "And it's not supposed to."

Deputy Sarah Odell greets us in the lobby. She is chipper but commanding, her dark hair pulled back into a tight bun, a ring of keys on one hip and a handgun on the other. After a round of introductions, she stows her weapon in a gun locker, and the double doors at the back

---

*In the 1980s, KMD designed Pelican Bay State Prison, the California supermax notorious for its harsh conditions and use of long-term solitary confinement. The firm's approach seems to have evolved with the times; in recent years, KMD has designed a number of more humane facilities, including juvenile detention complexes with plentiful natural light, multiple activity areas, and nature-filled campuses.

of the administration building unlock with a click. Odell escorts us onto the grassy campus. The sunlight is blinding. A vast lawn stretches out before us; a volleyball net is strung up over a patch of grass just to our right. A long sidewalk, bordered by thick-trunked palm trees, runs down the center of the forty-five-acre campus, from the administration building just behind us to the cafeteria and religious services building on the other side of the property. Classrooms, a library, and a recreation building all sit along this central promenade, while clusters of housing units, arrayed around several smaller courtyards, dot the campus perimeter.

Las Colinas, which has beds for about twelve hundred detainees, is home both to women who are awaiting trial and those who have already been convicted. All inmates receive a security classification, based on their charges and their behavior while incarcerated; about 70 percent of the women here are classified as a low-security risk, a designation that comes with a number of privileges, including the ability to walk around the campus without an escort, use the outdoor amenities, and visit the cafeteria for meals. (Low-security detainees who don't follow the rules can lose these privileges, while high-security detainees can gain them with good behavior.)

We follow the path to one of the low-security housing units. Three women in tan uniforms are talking quietly at a picnic table just outside the building. We walk through the doors into a large dayroom, a communal space where women can watch TV, play cards, read, or just hang out. The room holds several dozen light green armchairs and blond wood tables of varying shapes and sizes. Krueger and his colleagues had embraced an idea known as normalization, or the notion that correctional facilities should look less like impersonal institutions and more like actual homes. To that end, they tried to steer clear of the traditional, indestructible furniture and fixtures—steel toilets, concrete benches, metal tables bolted to the ground—that are so common in prisons.

Light pours in through the large windows, and a panoramic photo of the beach is mounted across the entire span of the back wall. Every housing unit has one of these murals, which depict landscapes from around San Diego County. (Several studies suggest that nature photos and videos can calm inmates and prison staff and reduce violent behavior.) Between the walls, which are painted in cool, calming shades of blue and green, and the blond wood accents, the unit looks like what might happen if IKEA designed a prison.

The living area, just beyond the dayroom, is entirely open plan, with no cells, no doors, and no locks. Each woman has what is essentially her own cubicle, containing a bed, a desk, a chair, and a small storage unit. Here, too, the furniture is a light seafoam green, with warm wood touches. The cubicles are stocked with personal effects—the one closest to us has a pack of playing cards, a sci-fi novel, a bottle of Dr Pepper, a brush, and a bottle of VO5 shampoo—but they're tidy, and, Krueger notices, there's not even the smallest speck of graffiti. "They keep it clean," Odell says. "They're taking care of it." The women are free to walk from their bunks to any of the unit's common spaces, including the communal bathrooms—which the women are delighted to discover have real mirrors—and a second, smaller dayroom toward the back of the first floor, which has a microwave, videoconference phones, and an air hockey table.

Instead of sitting in an enclosed guard station or office, the corrections officer on duty works from an open desk in the central dayroom. It's a departure from the way many correctional facilities are designed, with hard barriers—like bars or doors or bulletproof glass—physically separating guards from the inmates they're charged with overseeing. Here, the guards sit right out in the living area, where they can converse with detainees face-to-face.

The Federal Bureau of Prisons began experimenting with this arrangement, known as direct supervision, in the 1970s, and researchers have repeatedly found that it produces jails that are objectively safer, with lower rates of violence and assaults, than the classic model of indirect supervision. "The more contact that the staff had directly with the inmates, counterintuitively to some people, the safer everybody actually was," said Richard Wener, an environmental psychologist at New York University who conducted some of the first studies on direct supervision. "If the staff and the inmates were actually interacting, they started treating each other more like people instead of objects," he added. Facilities that use direct supervision also feel calmer and more relaxed, with lower levels of tension and stress for prisoners and officers alike.*

When corrections officers get to know incarcerated men and women on a personal level, they're better equipped to respond to their needs, Wener told me. That can be as simple as making sure that toilet paper is replaced promptly or as complex as helping inmates manage their emotions after a setback in court. Guards can help defuse tension as it builds, rather than waiting for a fight to break out and then rushing into the unit to respond. Under the direct supervision model, corrections staff are no longer mere security guards—instead, they're more like social workers, relying on communications, conflict resolution, and counseling skills.

Still, some corrections officials remain skeptical of the approach. Indeed, the Las Colinas deputies had initially worried that it wouldn't work, that the women would come up to the desk, mess with the computer and phone, and cause all manner of chaos. But those fears proved

---

*A 2014 study of inmates held at thirty-two different Dutch detention centers suggests that prison architecture can have a pronounced effect on inmate-staff relationships. Detainees who lived in campus-style facilities—in which the units are small and the officers tend to be close at hand—viewed their relationships with staff most positively, the researchers found. The inmates most pessimistic about their relationships with correctional officers were those who were held in panopticon-style buildings. The panopticon, a concept developed by the British philosopher Jeremy Bentham in the late eighteenth century, prioritizes surveillance and control. The buildings are round, with cells stacked in rings surrounding a central guard station. The design allows guards to keep tabs on inmates without the prisoners themselves knowing exactly when they're being observed. It also imposes a physical distance between inmates and guards, which may make prisoners feel more estranged from staff. Moreoever, the researchers write, "the large size and scale of the panopticon may increase anonymity and result in more impersonal and less frequent officer-prisoner interactions."

unfounded. "They respected the desk," Odell tells us. The women weren't interested in creating havoc, especially if they'd spent the day in a class or a vocational training program. When they got home, they just wanted to shower and relax, not cause trouble. "It took us a while to realize that," she says. That happened again and again, as the deputies realized that the women could be trusted with more freedom and autonomy than they'd first assumed.

The redesign, which increased the number of classrooms on campus from five to twenty, has made it possible for Las Colinas to vastly expand its programs and services. There's now a career center as well as several buildings dedicated entirely to vocational training, in fields ranging from landscaping to laundry. An enormous industrial kitchen houses a prized culinary arts program. Women in the six-month program, which follows a curriculum developed by the National Restaurant Association, prepare all the food for the staff cafeteria and operate the coffee cart that sits just outside it. "They learn how to make everything Starbucks makes," Odell tells us. Many of the program's graduates go on to work in local restaurants.

The medical clinic—which offers a full array of services, including dental care, obstetrical and gynecological care, and mental health treatment—is spacious, modern, and clean, with the same greens, blues, and blond wood tones that define so many of the jail's spaces. Once a year, a mobile mammography unit visits and offers free mammograms.

The campus is full of other nice touches. The family visitation room is open and bright, with floor-to-ceiling windows, a striped accent wall, and a crisp white-and-blue color scheme. Four private child visitation rooms are decorated in primary colors, covered in soft foam floors, and filled with books, toys, and plastic play equipment, like toddler-sized slides. An air-conditioned rec center is stocked with yoga mats and Zumba DVDs. And an outdoor amphitheater hosts holiday activities, Friday-night movies, and visits from a local symphony and theater company. "I don't feel like I work at a jail," Odell tells us. "At the old facility, I'm like, 'Heck yeah, I work at a jail.' It was sad."

IN 2015, CGL, the corrections consulting firm, conducted a post-occupancy evaluation, interviewing staff and detainees about Las Colinas. The inmates gave the design positive marks. "We only had one, out of the many that we interviewed, who said that she did not like the campus," Stephen Carter, the CGL executive, told me. "And when we pursued that a little bit, it was that she didn't like walking, and she had a long way to go between her housing unit and the dining hall, and she would rather have dined in her housing unit." The women liked the sense of community that the dormitory-style accommodations fostered and the direct supervision model. "For them, it just seemed to be natural that they could walk up to the correction officer and ask for something," Carter said.

The corrections officers were pleased, too. "Straight up and down, the staff said, 'It's a joy to come to work now,'" Carter said. Guards said that they felt less stressed, and nearly all of them preferred direct supervision to being tucked away in a control room.

The improvements in inmate behavior were noticeable: after the new facility opened, the number of assaults—inmate-on-inmate as well as inmate-on-staff—fell by 50 percent, according to data collected by the county. "It was almost immediate—you could see a difference," Brown-Taylor told me.

The results jibe with what Richard Wener has observed in his decades of research on jails and prisons. "One of the things that the physical environment does, that humans respond to, is that it gives you a set of expectations about how people are likely to behave and how people think you're going to behave—what we expect from you," he told me. "And when you go into a place that looks like the old Central Park Zoo, which is what a lot of these places were—the big bars, the concrete, and the floors that you could hose down, just like you do for the zoo animals—it tells you what they think about who inmates are. You know, 'These are dangerous animals that we need to keep locked up behind bars.'" A softer, more homelike environment sends a different message, Wener said. "It says, symbolically, 'We're thinking of you as being a civilized human being.'"

There were moments during my day at Las Colinas when I nearly forgot that I was touring a detention facility. For instance, the inmate cafeteria bore a surprisingly strong resemblance to the school cafeteria I'd seen in Buckingham County, Virginia. But there were sudden reminders that I was, in fact, at a jail: Three women standing against a wall, spread-eagled, being patted down by a deputy. High-pitched screaming coming from a padded cell in the intake area. There were design giveaways, too: the armchairs in the dayroom were filled with sand so that they couldn't be picked up and thrown, and the risers on the stairs were all made of mesh to ensure detainees couldn't hide beneath them.

And while I was genuinely impressed by the environment that the designers and the county had managed to create for its hundreds of low-security inmates, the approach only went so far. Women who were deemed a greater safety risk were assigned to high-security living units, which were sequestered at one edge of the property, behind a locked gate. There were no open-plan cubicles there; instead, the women lived in conventional cells and were supervised by teams of three guards who sat in enclosed guard stations. The mostly metal furniture was bolted to the floor. These women could use the dayrooms in their units and the exercise equipment in the adjoining rec yards, but they couldn't leave without an escort. "They're the least free," Krueger told me.

The units *were* an aesthetic step up from a conventional jail: they had the same panoramic nature murals as the lower security units, there was lots of natural light, and the cells had accent walls painted in pastel blues and greens. "We fought hard to keep that little punch of color," Krueger said. "We still tried to make it look nice. But it's a different kind of space."

Stephen Carter told me that, in retrospect, he's not sure they needed such stringent security measures. "If we were doing it again, we could go lighter on some of the security side of it," he said. "In my view, we probably could have done with fewer cells that are in that harder category."

There's been a lot of outside interest in Las Colinas; Odell has given tours to corrections officials and experts from as far away as Saudi Arabia and New Zealand. And a handful of similar complexes have sprung up across the United States. But the model isn't always an easy sell, especially to politicians worried about looking "soft" on crime. As Wener told me, "The easiest way for a sheriff to lose an election is for his or her opponent to say, 'He built and is running a country club for bad guys.'" Moreover, while criminal justice reform is gaining support, a substantial subset of Americans continue to believe that incarceration *should* be punitive and that retribution is more important than rehabilitation.

It's probably no accident that Las Colinas is a jail for women; the design team suspects that it might have gotten more public pushback if the facility had been intended for men. In fact, when New York City proposed replacing its notorious jail complex on Rikers Island, which primarily houses men, with several smaller, more humane jails in 2017, the *New York Post* proclaimed, "The mayor wants jails that feel like a retreat in Tulum." The tabloid illustrated the article with photos of Las Colinas.

Of course, a few nature murals and pastel armchairs do not a tropical resort make, and Wener believes that these kinds of critiques fundamentally misunderstand what incarceration is for. "The ultimate sanction is loss of freedom," he said. "People are sent to prison *as* punishment. They're not sent to prison *for* punishment."

Some progressives are deeply skeptical of the humane prison movement, too. To learn more about these critiques, I called Raphael Sperry, the president of Architects/Designers/Planners for Social Responsibility (ADPSR). Creating nicer prisons, he told me, lets us off the hook for larger problems with the American criminal justice system. In Sperry's mind, "The number one thing to understand about the system is that most people who are there are there unjustly. You can't make up for a lack of freedom by putting them in a really, really nice box," he told me.

The priority, Sperry said, should not be building better prisons, but reducing the number of people who are incarcerated in the first place. Architects who want to be part of the solution should focus on designing more affordable housing, drug treatment centers, mental health facilities, and all the other buildings that are part of a social safety net. "I'm not against designing for health and wellness," he said. "I'm all about that. I just think that if you want to heal the kinds of problems that lead to crime, you don't do it in a prison—you do it in a neighborhood."

I agree with Sperry. We need to invest in communities and neighborhoods and public health and education; in restorative justice and other alternatives to incarceration; in bringing the prison population down. And yet, even if we accomplish all of these things, prisons won't disappear, and there is value in minimizing the harms they cause.* These goals aren't contradictory. We should send fewer people to prison, and we should treat them better while they're there. (And to be clear, treating them better isn't only a matter of picking calming paint colors. It also means curbing abuse and ensuring that inmates have access to basic necessities, including adequate food, sanitation, and medical care.)

In the years to come, we should also pay close attention to the effects that these reforms are actually having; after all, there's a long history of trying to remake American prisons, and those efforts haven't always worked out well for inmates. It's not yet clear whether more humane prisons improve long-term outcomes, but Las Colinas is beginning to track how many of its former detainees end up returning to

---

*There is a prison abolition movement whose members argue that we should not build any new correctional facilities and should gradually decommission those that already exist.

custody. "These folks come back to society, which I think people forget sometimes," Krueger told me. In the United States, ten thousand people are released from prison every week. "And how they're treated while they're in there has a lot to do with how they're going to act when they get out."*

The bottom line of the humane prison movement, Wener told me, is something your grandmother might have taught you: "Ultimately, if you treat people better, they respond better."

THAT'S TRUE OUTSIDE of prisons, too, and designers of other kinds of residential facilities, from nursing homes to mental hospitals, have begun to embrace the concept of humane design. According to Francis Pitts, a New York–based architect who specializes in psychiatric facilities, the move is long overdue. "Most traditional mental hospitals are not designed for the human experience," said Pitts. "They are soulless, bureaucratic, processing machines." In fact, he told me, many bear a resemblance to prisons—cold, institutional, and designed primarily to keep residents under control. "They were designing for the illness and not for the people who had the illness, and it was wrongheaded," he said.

Psychiatric hospitals are increasingly trying to create warm, aesthetically appealing environments that make patients feel more at home. They are giving patients the ability to personalize their spaces and using real dishes and glassware for meals. And they are employing many of the same strategies that other health-care, educational, and residential designers are embracing, from adding rooftop gardens to creating small group spaces that promote a sense of community. In addition to improving patient well-being, these kinds of thoughtful de-

---

*If the humanitarian argument doesn't convince you, consider that there's a compelling economic case for helping former inmates rebuild their lives and stay out of the correctional system: mass incarceration is *expensive*. "If we don't reduce the recidivism rate in this country, then we're going to bankrupt states and counties, especially local governments," Carter told me.

sign touches, Pitts told me, send patients an invaluable message: "That they are in a building, in a place, that offers dignity and hope."

Indeed, the physical environment can send a powerful signal about who, and what, we value. In a 2012 study, Lorraine Maxwell, an environmental psychologist at Cornell University, found that adolescents' perceptions of themselves were tied to the quality of their school buildings. Students who attended class in high-quality buildings—those that were, for instance, clean, bright, well ventilated, and well maintained—not only performed better academically but also were more confident in their own ability to succeed than those who spent their days in dark, dirty, run-down buildings. A "good" school building, Maxwell theorizes, sends students the message that they matter; a "bad" one does the opposite.

That's true at larger scales, too. People who live in litter-strewn neighborhoods report having less community pride and less trust in local government than those who live in tidier ones. And though urban green spaces have numerous benefits, including bolstering mental health, unkempt landscaping can backfire; people who live on blocks on which the greenery is poorly cared for have lower levels of civic trust than those who reside on blocks that don't have any vegetation at all. Sprucing up vacant lots, on the other hand, appears to reduce some kinds of crime and resident stress, improve perceptions of public safety, and draw residents out of their homes, making them more likely to socialize outside.

The studies are part of a growing field of research—promoted by organizations like Happy City in Vancouver and the Centre for Urban Design and Mental Health in London—that suggests how we might build cities that nurture mental health. Good urban design, some say, might even help us fight the so-called epidemic of loneliness. "There's good evidence that loneliness and social isolation seems to be increasing around the world, especially in urban areas," said Kevin Bennett, a psychologist at Pennsylvania State University's Beaver campus. "Even though people are living in a small geographic space, densely populated, millions of people, the rates of isolation are incredibly high."

Although urban living is a far cry from life in prison, Bennett sees clear parallels between the two environments; some of the same psychiatric conditions that plague inmates in solitary are especially prevalent in cities, he noted. "We see similar effects, although not as extreme, in urban environments where people feel isolated, they feel like they're living in a small apartment where they're disconnected from other people," he explained.

Cities haven't been deliberately designed to isolate people, but their scale and density can be overwhelming, the sheer number of unfamiliar faces preventing residents from developing strong bonds with any one of them. In a 2012 survey of nearly four thousand people, the Vancouver Foundation found that people who lived in high-rises were more likely to say that they never chatted with their neighbors than those who lived in town houses or single-family homes. Only 56 percent knew the names of at least two of their neighbors, compared to 81 percent of those who lived in detached homes. High-rise residents were also less likely to report trusting their neighbors or doing small favors for them and more likely to say that they were lonely and had trouble making new friends.

But there may be ways for designers and urban planners to help foster more social interaction. "Let's say you have a public space—indoor or outdoor—just the way you set up the benches or the tables can encourage a certain amount of communication," Bennett told me. Studies of lounges in long-term health-care facilities have shown that arranging chairs in small, intimate clusters, rather than in rows around the room's perimeter, can encourage socializing. Urban designers could borrow these lessons, designing small groupings of outdoor furniture and positioning park benches so that they face each other. "You don't expect people to get into life-changing conversations," Bennett said, "but they get to interact for two minutes and then they go on to the next part of the day and they might get to interact with someone else for an extra two minutes, and it's sort of the cumulative effect of all those interactions that I think is helpful."

Researchers have also found that people who live in neighborhoods that are chock-full of amenities—including parks, libraries, and restaurants—are more likely to socialize with their neighbors and feel less isolated than those who live in less vibrant areas. At the building level, providing apartment residents with shared spaces, like playrooms and community gardens, could serve the same purpose, allowing people with common interests to gather.

Finally, promoting urban well-being requires paying attention to the fundamentals, like ensuring that housing is safe, clean, quiet, and well maintained. Affordability is essential; in one large, longitudinal study, researchers found that when low- and moderate-income Australians moved out of affordable housing and into homes that stretched their budgets, their mental health suffered.

For as grand as the goals of humane design can seem, the animating impulse is simple. "At the heart of the human-centered [design] movement is a commitment to kindness," Pitts told me. "Design can definitely be a way to be kinder."

Or smarter. Just as our evolving societal values are prompting us to create spaces that are kinder, advances in technology are helping us engineer buildings that are more intelligent. By now, it's clear that even the smallest-seeming decisions about the indoor environment—whether to open a window, where to place the stairs, how to arrange the furniture—can have significant ripple effects on our lives. Smart home sensors and systems are imbuing our buildings with new powers, including the ability to play a more active role in our welfare.

SEVEN

IF THESE WALLS COULD TALK, LISTEN, AND RECORD

F OR MORE THAN half a century, writers and futurists, from Ray Bradbury to the creators of *The Jetsons*, have been conjuring up intelligent, high-tech homes that essentially run themselves. In their versions of the future, our homes would be much more than shelter: the houses of tomorrow would cook, clean, and care for us. They'd wake us up, make us breakfast, and then tidy up afterward.

Tomorrow is here. In homes across the world, smart thermostats glow, autonomous vacuums spin, and intelligent speakers stand at attention. Programmable shades rise with the sun, and connected refrigerators monitor our supply of milk. We can rely on smart flowerpots to water the plants, smart pet feeders to dispense kibble to the dog, and smart locks to let the maintenance worker in—all while we're out of the house. And at the end of a long workday, we can slide dinner into the oven, which will detect that we're cooking a chicken and then roast it to perfection, and then collapse atop a prewarmed mattress, which will dynamically adjust its temperature and firmness throughout the night. As morning approaches, the mattress will gradually cool to help rouse us out of our slumber.

"This area moves pretty fast now," said Juan Carlos Augusto, who heads the research group on the development of intelligent environments at Middlesex University in the United Kingdom. "Many things that were thought to be borderline science fiction a few decades ago are starting to hit the market." Tech companies are investing heavily in smart home systems—also known as intelligent environments, ambient intelligence, home automation, pervasive computing, and the internet of things—and consumer appetite is growing. By 2023, more than half of American households, and one-sixth of those around the world, are expected to have smart home devices.

The first wave of products aimed to alleviate the hassles of daily living, tracking our behaviors and preferences in order to make our homes more comfortable, convenient, and efficient. But now Big Brother is getting an MD. The smart mattresses that keep us comfortable are also collecting reams of data about our sleep quality, heart rate, and respiration, while some smart thermostats are monitoring the air quality in our homes. Many companies now sell smart pill bottles that light up, chime, or fire off text messages when patients miss a scheduled dose. Google has patented an optical sensor that would enable smart mirrors to monitor our cardiovascular health by detecting subtle changes in skin color, and Amazon has patented a system that could prompt our smart speakers to order cough drops when they overhear a sniffle. (It would, presumably, order those lozenges from Amazon.)

These advances, and the ones still to come, mean that our homes are becoming more intimately involved in our health care than ever before. Even as these technologies move into the mainstream, there's one specific segment of the market that's ahead of the curve, and it isn't the one you might expect. It's not the tech-savvy millennials who are pioneering the most advanced versions of these products—it's senior citizens. It's in the senior care community where scientists and engineers are putting all the pieces together, experimenting with integrated health-monitoring systems that are truly turning homes into medical devices.

The potential upside is enormous. There are more than 700 million people on the planet who are sixty-five or older, a figure that is expected to double by 2050. Moreover, our lengthening life spans mean that seniors are living *longer*, spending more years in a phase of life that is often accompanied by increasing frailty and diminishing cognitive function. In theory, at least, in-home sensors, cameras, trackers, and monitors could help keep many of these older adults healthy and independent, allowing them to age safely in their homes, even in the face of illness and infirmity. These systems provide a glimpse of the future of smart home health monitoring—a preview of both the promise and the risks of letting our buildings play doctor.

MARILYN RANTZ, a gerontological nurse and researcher at the University of Missouri, has dedicated her career to improving the experience of aging in America. Typically, as seniors' needs increase, they get shuttled off to a series of residences that provide escalating levels of care—from their own houses to assisted living facilities to nursing homes. These moves, however, can be disruptive and dangerous. Transferring elderly men and women from one home to another, especially against their will, can cause confusion, sleeplessness, loss of appetite, anxiety, and depression.

It can have more serious consequences, too, as Rantz herself discovered in the 1980s. At the time, she was an administrator at a nursing home in Wisconsin; after the staff moved a group of its residents to a new unit, mortality rates spiked. "Those moves are very difficult for the person to adjust to, create a lot of additional stress, and in fact can precipitate an early demise," said Rantz. Seniors aren't too keen on them either; in survey after survey, they report that they want to stay in their homes and remain independent as long as they possibly can.

In 1996, Rantz and some of her colleagues at the University of Missouri's Sinclair School of Nursing decided to establish a new model for senior housing that would allow people to "age in place." They partnered with the Americare Corporation, which owns and operates senior living facilities, to create what they came to call TigerPlace, a nod to the University of Missouri's feline mascot, and launched a health-care agency, which they called Sinclair Home Care, to administer it. TigerPlace would be a complex of apartments where adults could live out the entirety of their golden years. It would be an independent living facility, but an adaptable one. If a resident's needs changed, Sinclair could bring in additional aid, like physical therapy, wound care, assistance with bathing and dressing, and, when the time came, hospice care. "We bring services to people and help them age there through the end of life," said Rantz.

In late 1999, while Rantz was in the midst of trying to get TigerPlace up and running, her elderly mother fell and shattered her shoulder. She was home alone and lay on the floor for eight hours before someone came to her aid. She never really recovered. "She was dead well within six months of that fall," Rantz told me. "It was a classic example of what happens to people after falls and fractures."

More than one in four older adults fall every year, and those who spend long periods of time lying on the floor have an especially poor prognosis. Rantz wondered if she and her colleagues could devise a technological system that would automatically detect falls at Tiger-Place so that no one would have to suffer her mother's fate.

Searching for collaborators, Rantz walked across campus to the College of Engineering, where she gave a talk about what she envisioned. Marjorie Skubic, a professor in the department of electrical engineering and computer science, was in the audience. Skubic had arrived at the university in 1997 knowing that she wanted to work with academics in other disciplines and do research that made a difference. ("I didn't want to just write a bunch of papers that would appear in some stuffy journal that maybe ten people would read," she told me.) Rantz's vision ticked both of those boxes. "I said, 'Wow, this is exactly what I'm looking for,'" Skubic recalled. "I raised my hand."

In a series of focus groups, Skubic and Rantz learned that while seniors were open to the idea of fall detection technology, they were wary of wearables, like pendants, bracelets, and clip-on devices. These gadgets require the user to do a lot of work—chiefly, to remember to wear them at all times and to keep them charged—and they can stigmatize the wearer as ill or frail. So Skubic and Rantz decided to focus on developing sensors that could be deployed in seniors' homes. "We said, 'Let's just make it so they don't have to do anything,'" said Skubic. "You mount them in the environment, there's nothing for them to wear, you don't have to charge anything."

Working with several colleagues and a large team of students, Skubic spent several years playing around with fall detection technologies in her cluttered university lab, which is stocked with high-tech motion-capture cameras and a haphazard assortment of household furniture. They tested floor vibration sensors, Doppler radar, and microphone arrays designed to capture the sound of falling before deciding that the most promising method relied on a piece of video game technology: the Microsoft Kinect, a now-discontinued motion-sensing camera that worked with the company's Xbox gaming consoles. The Kinect contained a depth sensor that measured how far away objects were from the device. From this depth data, Skubic and her team could extract the silhouettes of human figures, track their movements over time, and detect when they took a tumble.

To test and refine the system, Skubic's team hired stunt actors and asked them to perform twenty different types of falls, including tripping, slipping, falling out of a chair, and rolling out of bed. "We trained the stunt actors how to fall like elderly people," Skubic explained. They also asked the actors to perform movements that might trigger false alarms, like picking up an object from the floor, and used machine learning to train the software to distinguish between the real falls and the impostors. "The key point about fall detection is not just detecting the falls per se, but it's distinguishing the falls from the non-falls," Skubic told me. When they were satisfied with what they'd built, they took it to TigerPlace.

I VISITED TIGERPLACE in the summer of 2018. The sprawling, one-story brick building, which opened in 2004, is located just a few miles from the University of Missouri campus and has a lot of luxe touches: a grand piano in the lobby, brocade curtains, and so many chandeliers that I lost count. Skylights bring sun into the interior corridors, and residents have easy access to the outdoors. Every apartment has its own screened-in porch, which opens directly to the outside, and a large communal courtyard features fountains, bird feeders, and pots filled with herbs. (The outdoor access is especially useful because TigerPlace is pet-friendly. "More than just pet-friendly," said Kari Lane, the director of operations at TigerPlace. "Pet-encouraged." The residence is typically home to more than a dozen dogs and cats and even contains a veterinary office, which is staffed by faculty and students from the university's veterinary school.)

The residents range in age from their mid-sixties to their late nineties, and the majority of them have at least one chronic disease. There are seniors with heart disease, arthritis, diabetes, and dementia; the typical tenant takes more than a dozen medications. They get comprehensive health exams every six months, though they retain their own personal doctors, who stay in close touch with TigerPlace nurses.

The goal is to keep seniors healthy, of course, but also to enable them to continue living in their homes, even as their ailments wax and wane. "So if someone's having trouble walking, we might call in a physical therapist," Lane said. "And when they don't need that anymore, we pull that back. If they fall and break a hip, they would go to the hospital, maybe go to rehab like anyone else would, but then they would come back to their home."

The approach seems to be working; the average length of stay has been increasing since TigerPlace opened and has surpassed the average tenure at senior living facilities nationwide. Some residents stay for more than a decade, all the way up through their very last hours of life. On the day of my visit, a man who'd moved into his apartment fourteen years prior was dying. "We don't think he'll make it through the day," Rantz told me. But even as the staff struggled to get their tears under control, they recognized that, in many ways, the story was a happy one. An elderly man had spent his final years living in his own home, where he was being shepherded to a peaceful death.

Researchers from the University of Missouri are constantly testing out new ideas at TigerPlace, and the bulletin board outside the dining room is papered with flyers advertising studies that need volunteers. Skubic's team launched a fall detection study in 2013, installing the depth-sensing cameras in a dozen TigerPlace apartments. Each time the system registered what it thought was a fall, it fired off an e-mail to the nurses on duty. Each e-mail contained a short video clip of the movement that triggered the alarm, allowing the nurses to quickly see exactly what had happened. If the video showed a real fall, the nurses could be in the apartment in a matter of minutes. If it was a false alarm—just a dog jumping off a couch or a grandchild throwing himself to the floor, for instance—the nurses could stand down. "If they didn't fall, I don't have fifteen people run into their room to check on 'em," Lane said. "We want our residents to have as much privacy and dignity as possible. We try to prevent those types of invasions."

Over a period of 101 days, the system detected about 75 percent of falls, generating one false alarm per month per apartment. The system proved popular, and the research team subsequently deployed it in additional apartments at TigerPlace and at several other senior living facilities throughout the state. (The algorithms also got more accurate over time and now detect nearly all falls that are within the camera's field of vision, Skubic told me.)

But from the beginning, the researchers' goals had been much grander than simply responding to crises. Sensors installed in the home, they thought, offered them an opportunity to detect subtle changes in seniors' health and behavior and to diagnose small problems before they ballooned into big ones. If they could identify folks at *risk* of falling, they could send in physical therapists to help them improve their strength and their balance; if they suspected that residents were developing pneumonia, they could start them on antibiotics and fluids.

Previous research had shown that seniors who walk slowly and take short strides are especially likely to fall, so Skubic and her colleagues developed gait-monitoring software for depth-sensing cameras. When they tested it at TigerPlace, they found that even slight declines in walking speed and stride length were predictive of falls. The staff now gets alerts if the system calculates that someone's odds of falling have risen above 85 percent. It's the kind of system that Rantz wishes had existed two decades ago, before her mother fell. "We could've known that that was going to happen to my mom three to four weeks ahead, and we could have gotten her treatment," she told me.

Skubic's team also developed a hydraulic bed sensor, an array of long, flexible, water-filled tubes that sits under a mattress. The sensor monitors nighttime restlessness and measures heart rate and respiration by picking up on the minute movements of the heart and lungs. In addition, simple motion sensors mounted in the apartments provide clues about tenants' daily movements and routines, tracking each time they enter and exit a room, for instance.

Together, these sensors perform what amounts to a constant virtual checkup, notifying the nurses if they detect any worrisome behavioral changes. A sudden uptick in nighttime restlessness might indicate pain; more trips to the bathroom might be an early sign of a urinary tract infection. A woman who's spending more time in bed might be fatigued or depressed; one who begins leaving her apartment in the middle of the night might be in the early stages of dementia.

When the nurses receive an alert, they go chat with the patient in question. "We don't go in and say, 'Hey, you had an alert last night,'" Lane said. "But we go in and use careful questioning and relationship building to figure out how they are feeling and determine if anything is going on that we need to pay attention to. It's kind of like detective work."

In one recent case, the staff got an alert that one woman had been more restless than usual at night. When a nurse went to check on her, she admitted that she'd been sleeping poorly—and that she'd developed a funny tingling in her hands. That set off alarm bells; tingling hands can be a sign of dehydration and an electrolyte imbalance. The nurse called the woman's doctor, who ordered lab work that confirmed the diagnosis. "And they were able to push fluids to avoid a hospitalization," Lane said. Over the course of a single nursing shift, the woman had been evaluated, diagnosed, and treated—all without leaving home.

In another instance, the staff received an alert that one man had begun wandering around his apartment in the middle of the night. They discovered that he was in the early stages of dementia and started him on therapy and medication designed to slow its progression. "I could give you a ton more examples," Lane said. "How many do you want?"

The system has helped the nurses catch pneumonia, delirium, heart failure, hypoglycemia, and other problems. It can detect changes weeks before significant health events, like falls, ER visits, and hospitalizations—before people report symptoms to nurses and, sometimes, before they even perceive them. In a 2015 study, the research team found that TigerPlace residents who'd had the sensors installed were able to live in their apartments for 1.7 years longer than those who'd chosen to forgo them.

TIGERPLACE HAS PROVIDED powerful proof of concept, and the team is working to build upon what they accomplished there. One of Skubic's former graduate students has formed a company called Foresite Healthcare to commercialize the technology. Foresite has developed its own depth-sensing camera to replace the Microsoft Kinect and has managed to bring the fall detection false alarm rate down even further, to one every three months. The company has installed its system in dozens of senior living facilities in multiple states.*

Skubic and Rantz are thinking through how they might adapt the illness detection system for private homes. It's not as easy to translate as it might seem, a lesson that Skubic learned firsthand when she installed one of the sensor systems in her parents' South Dakota home. Shortly thereafter, her ninety-six-year-old father was diagnosed with pneumonia. When Skubic found out, she went back and looked at his data. The bed sensors made it clear that, over the previous six weeks, her father had become more restless. In fact, his restlessness had been significant enough that Skubic had received alerts about it. But she didn't know how to interpret them—or what to do in response.

If her father had lived at TigerPlace, a nurse would have reviewed the alert against his medical chart, compared it to similar alerts she'd received in the past, and then examined him in person. But Skubic lived across the country from her father, and she was an engineer, not

---

*Foresite has also developed a companion system to help prevent patient falls and bedsores in hospitals. The sensors detect when patients are moving in ways that suggest they're preparing to get out of bed so that nurses can come and assist them. In addition, they notify nurses when bed-bound patients need to be repositioned in order to alleviate the prolonged pressure that can cause bedsores.

a nurse. Restlessness itself didn't necessarily portend a problem—there are lots of reasons someone might not be sleeping well—so she had just filed it away as an anomaly.

The only reason Skubic's father ended up seeing a doctor was that his wife of more than seventy years insisted—she knew that he didn't seem like his usual self. "So she basically was doing what we want our sensor system to do," Skubic said, laughing.

The experience was instructive. "I mean, I've been working in this for ten, actually fifteen years, and I didn't know how to interpret the alerts," Skubic told me. "We need to a do a lot better job of turning all this data into useful information, and we need to really make it simple."

Skubic and her colleagues are working to translate their wonky, data-laden alerts into plain English. They're exploring the possibility of delivering them through commercially available smart speakers, like the Amazon Echo or Google Home. "It's set up as a conversational system so that you can ask questions like 'How did I sleep last night?' or 'What is my fall risk?'" Skubic told me. "It also will support the use of family members, so that a family member could say 'How did my mother sleep last night?' Or 'Does my mother have fall risk?'"

The bigger challenge is figuring out what comes next. At Tiger-Place, the pathway for the nurses is clear. But what should an elderly woman do if she's living on her own, in her own home, and Alexa tells her that she hasn't been sleeping well lately? "The question becomes 'Is it bad enough that I need to go see a doctor?'" Skubic said. "'Or is this just something where I need to change what I eat or I need to go to bed earlier at night? Or am I taking the wrong medications?' So somebody has to be able to look at this in a holistic way and give people some guidance." Routing all of the alerts to the relevant physicians doesn't seem like a reasonable solution, given how overloaded most doctors already are, so the team has been contemplating other solutions, like creating a network of nurses or "care coordinators" who can help consumers review and interpret their alerts.

While many seniors are legitimately excited about these technologies, some have expressed concerns that the devices will be intrusive, unreliable, or downright anxiety-provoking. Not everyone wants to see the excruciating minutiae of their body's inner workings charted out for them in a daily e-mail. "Some of them have said, 'I don't want to see that, I don't want to be a hypochondriac, I don't need to know that,'" Lane told me. "'Someone'll let me know if I need to go to the doctor.'"

Even those who are open to the technology have concerns about privacy. "They would say things like, 'I'm okay if my daughter sees it, but I don't want my son to see it because I don't trust my son's wife,'" Skubic told me. "They viewed it as their data, which I thought was an interesting perspective and actually an encouraging perspective." And although the fall detection cameras show the body only in silhouette, some seniors have asked not to have them installed in their bedrooms or bathrooms.

That's understandable, but it comes with trade-offs. The more that developers limit the scope of the surveillance, the more the system will miss. That caused some misunderstandings when the TigerPlace researchers installed a few of their sensor systems in private homes on a trial basis. "Someone would call and say, 'Well, I had a fall, and you guys didn't call me,'" Lane told me. "Well, that fall wasn't where the sensors are installed. Because they tried to install the fall sensors where the person felt they were most at risk for falling and also where the person was okay with them being installed for privacy reasons."

Likewise, other senior care facilities that are interested in adopting the technology don't always understand its limitations. "They want it to be plug and play," Lane said. But getting the most out of it requires staff education and, often, a culture change, a commitment to being less reactive and more proactive. After all, the technology isn't magic, and it's not the sensors themselves that are keeping TigerPlace residents healthy. "The technology that we use here is wonderful, but it still needs nurses behind it," Lane said. "It's just another tool, like a stethoscope, to help us make informed choices."

THE TOOLBOX IS getting bigger as engineers all over the world develop health-monitoring gizmos and gadgets. A group of researchers from several Japanese universities built a bathtub that passively and painlessly measures the electrical activity of the bather's heart, potentially detecting heart disease, arrhythmias, and other conditions. A team of MIT engineers developed a device, which can be hidden behind a wall, that tracks heart rate and breathing using wireless signals that detect tiny chest movements. There are motion-detector-based systems that keep tabs on whether seniors are taking their meds and eating regular meals, and intelligent assistants that remind adults with dementia to shut the front door, turn off the tap, and close the windows when it rains. There are even robots that work in concert with home health sensors; they prompt people to drink when they're dehydrated, and they scoot over to people when they fall. (The fall-response robot then messages a caregiver, who can use the machine to video chat with the resident.) Elder care robots have become a focus of intense research and development in Japan, where the population is aging especially quickly.

Some research teams are deploying smart home technologies to help people with other disabilities and conditions. (Recall, for instance, that First Place installed a motion-sensor-based system to help ensure that its neurodiverse residents don't accidentally leave their stoves on.) The U.S. Department of Veterans Affairs has developed what it calls a "home-based cognitive prosthetic" to support veterans with traumatic brain injuries, which can impair memory, planning, and problem solving. The system tracks the locations and activities of the vets as they move around their homes and employs wall-mounted touch screens to display customized reminders and prompts—to make breakfast, brush their teeth, take their medications, or empty the washing machine.

As computing power and internet speeds ratchet up, the possibilities multiply, especially as smart cities begin collecting unprecedented new volumes of data. Cities are experimenting with electrical grids that respond to changing power demands; streetlights that dim when foot traffic is low; and sensors that signal when parking spaces are free, garbage is ready for collection, and roads require repair. "Big data is like a new natural resource, and it's as impactful on how you organize cities as water and electricity were one hundred years ago," said Joe Colistra, who directs the Institute for Smart Cities at the University of Kansas.

An intelligent metropolis could also provide new ways to monitor and manage public health, Colistra told me. In the spring of 2018, he invited me to Lawrence, Kansas, to see what he had in mind. We walked to the back corner of a cavernous, sawdust-scented warehouse, where Colistra had erected the wooden frame of a small house. It was an early prototype of his vision: a modular, affordable, prefabricated housing unit with health sensors built into its bones. These units, he said, could be installed in cities by the thousands.

Colistra was in the middle of testing out his fall detection and gait analysis system, which relies on sensors installed under the floorboards. It was cold in the warehouse, but Colistra took off his houndstooth sport coat and draped it gently over a pile of wooden planks. "We're gonna have to get dusty," he told me. We dropped to the ground and slithered under the floorboards. Rows of strain gauges, which detect force applied to the floor, and accelerometers, which can measure its vibrations, were attached to the wooden joists. "We recalibrated them to pick up heel strike," Colistra said. "So as you walk across the floor, we can see patterns of how people are walking."

The data captured by the sensors, which take two hundred readings a second, is sent wirelessly to the cloud. Colistra is working with mathematicians and clinicians to see if they can develop algorithms that detect not just falls but also limps, tremors, Parkinson's, flare-ups of multiple sclerosis, and even diabetes, which can cause a deadening of the nerves in the feet, leading to subtle changes in gait.

The wired-up floor is the first piece of what Colistra envisions as a much bigger system. He wants to use smart mirrors to monitor moles, dental plaque, eye reflexes, and the slight facial asymmetries that could signal a stroke.* And he's especially excited about the prospect of using toilet sensors to detect dehydration, kidney disease, and diabetes. He thinks a smart, hydration-sensing toilet could be a valuable tool for people with heart failure, who must take precisely calibrated doses of diuretics to flush their bodies of excess fluid. "The data from the toilet could be coupled with an automated medicine dispenser to adjust diuretic on the fly," Colistra said. "The [housing] unit becomes like a medical device. It's taking care of you."

Colistra's dream is to combine the information that these at-home gadgets are harvesting with all the other data that smart cities, and the companies catering to them, are collecting. "If you knew that someone had only four hours of sleep for two nights in a row, and they were beginning to limp, and the mirror told you that their eye tracking was off and their reaction time was slow, and the toilet told you they were dehydrated, and you overlay that with other environmental conditions, like it's high humidity, it's close to freezing temperature outside, there's a little bit of rain on the sidewalk, and there's a little bit of ice forming—you could start to predict with great accuracy, maybe that person has a ninety-nine percent chance of falling the next day," he told me. "So the power, then, is if you had these sensors in a neighborhood of ten thousand and you were able to predict some tiny fraction of the population that had a ninety-nine percent chance of falling, you could contact those ten or twenty people or their families and say, 'Hey, you better take a little extra care tomorrow or get a ride to the grocery store or have the manager of the assisted living facility check in on them.'"

According to Colistra, this flood of home health data could help city officials track the well-being of the population at large and identify

---

*There's a lot of ongoing research in this area. Optical sensors and algorithms can detect skin cancer; anemia; elevated cholesterol, blood sugar, and heart rate; stress, fatigue, and anxiety; and a variety of rare genetic diseases by scanning people's eyes, skin, and face.

areas where certain public health interventions, including active design strategies, might do the most good. He gives me an example: Say that readings from the smart floors and toilets reveal diabetes clusters, neighborhoods where people are developing the disease at sky-high rates. Researchers and officials could then work together to identify—and remedy—some of the causes. Maybe these neighborhoods need more parks or walking paths or affordable grocery stores. "It could have broader implications for how you plan cities or even in an ideal world, how you would prioritize money to certain neighborhoods," Colistra said. "If you actually use the data where it's all talking together, it's really transformational."

WHAT MAKES THESE systems so powerful, however, also makes them dangerous. They hoover up vast amounts of data about our bodies and the things that we do in the privacy of our own homes.* Smart home devices are tempting targets for hackers, and even when companies are careful, leaks and breaches are inevitable. It's only a matter of time before some users, at least, find their personal health data exposed to the world.

Beyond that, one of the real existential risks of smart homes and cities is that we end up turning our public and private spaces over to corporations, which can use our personal data for their own gain. It's already happened. In October 2018, *The New York Times* reported that a smart thermometer start-up had sold its customers' health information to Clorox, which used the data to ramp up advertising for its disinfecting products in areas of the country where there had been a spike in fevers. A month later, ProPublica revealed that internet-connected CPAP machines, which help people with sleep apnea breathe at night,

---

*One example of just how personal this data can get: In 2017, a Canadian company settled a class-action lawsuit over claims that it was surreptitiously tracking how individual customers were using its Bluetooth-connected vibrators. The company claimed it needed the data for "market research."

often share data with patients' health insurance companies; if patients don't use the machines reliably and correctly, the insurers can refuse to cover their cost. (Some digital health companies disclose the fact that they're sharing our data in impenetrable terms of service agreements, while others don't disclose it at all.)

Most troubling is the possibility that these products could be used to coerce patients into undergoing unwanted medical treatments. Some psychiatrists, for instance, have touted smart pill bottles as a way to help ensure that patients with bipolar disorder and schizophrenia take their meds as prescribed.* But what if they don't? Will doctors drop them? Will insurance companies jack up their rates?

Even if we give permission for a company to collect our health data, it's hard to predict how that information might be used in the future. Data analysis and machine learning are rapidly growing more sophisticated, and in a few years' time, researchers, doctors, and insurance companies may be able to pull entirely new insights out of the health data we're giving up today. In other words, we could end up exposing far more about ourselves than we've bargained for. Our health data could also reveal personal health information about genetic relatives who never gave their consent in the first place. (The question of consent becomes even thornier when considering technologies designed to monitor the elderly, who may have dementia or other cognitive disabilities.) And if these systems and products come preinstalled in our houses, apartment buildings, hotels, and hospital rooms—this is, in fact, already beginning to happen—will any of us truly be able to opt out?

Beyond these risks, not everyone will benefit equally from these technologies, which can be expensive to buy, install, and maintain. If these systems are truly transformational, they could widen the health disparities between well-educated, high-income consumers and everyone

---

*The effectiveness of electronic pill bottles remains unclear. In one large, randomized, controlled trial, researchers found that the technology did not improve medication adherence in patients who had recently suffered from heart attacks.

else. These inequities could be exacerbated by biases in the systems themselves, many of which are based on data that comes from healthy, able-bodied white men. Some facial recognition software, for instance, is much more accurate for light-skinned men than for women or people with darker skin tones. It's not hard to imagine that the same might be true of intelligent health-monitoring systems, especially given how much of our medical knowledge comes from studies of white men. (Skubic has also noticed that smart speakers understand the voices of young adults better than those of their elders.)

Then there are the technical failures. Malfunctions are annoying when your smart thermostat gets the temperature wrong, but when smart home gadgets become medical devices, failures can be catastrophic. "Imagine all the potential for things that could go wrong," said Juan Carlos Augusto, the intelligent environment researcher at Middlesex University. "It could be that it misses one important clue, and that's the one important opportunity where this person is in trouble."

In addition, constantly monitoring otherwise healthy people for a grab bag of different diseases can do more harm than good. False positives can be intrusive and undermine trust in the system, but they can also have more serious consequences. If a smart bathtub incorrectly alerts a bather to an abnormal heart rhythm or a smart mirror classifies a normal mole as worrisome, it could send perfectly healthy people to doctors for costly and time-consuming tests, some of which come with health risks of their own.

I also wonder about the social consequences of outsourcing more and more tasks to ever-smarter machines. Our relationships with other people are key to our health and happiness; we know that many seniors suffer from loneliness and that strong social networks are key to helping them age in place. But what if smart home products push human caregivers out of the picture? "We don't pretend to think that this technology is like a silver bullet," Colistra told me. "We still think the most important part of health and wellness is social connectivity, so having neighborhoods where people can support one another is probably way, way more beneficial than actually picking up data on someone's urine or their gait."

Likewise, technology that is designed to help increase independence might ultimately end up compromising it. Door alarms and GPS trackers may help keep seniors with dementia safe, but they also infringe upon their autonomy. It's going to be tricky to strike the right balance—and to determine when these kinds of privacy invasions are warranted.

For his part, Augusto thinks that a lot of the products that have been developed for the senior care market are "dehumanizing," policing people's behavior in their own homes and informing others of their failings and weaknesses. That's why he's deliberately taking a different approach. He's developing a system, for adults in the early stages of senility, that will act less like a snitch and more like a coach. It will alert users directly when it notices that they've been sleeping poorly or skipping showers or missing meals, changes that they might not be consciously aware of. "It's like a friend that reminds them of things or notices some deviations that are not that healthy," Augusto explained. The software will then give seniors advice about how to improve their sleeping or eating habits or send them reminders to shower and eat. "We are trying to create a more friendly, more trusting relationship between technology and the user," he said.

Augusto has proposed a set of ethical guidelines that he hopes developers will use when they're creating smart home systems. Intelligent environments, he says, should "actively benefit" users, be designed around what they actually want and need, respect their autonomy and dignity, and be accessible, affordable, and usable to people of all backgrounds and abilities. The technology should be reliable, stable, and secure, and it shouldn't displace human caregivers. Developers should be transparent with users about the extent of the monitoring and data collection; how the data is being used; and the risks, vulnerabilities, and limitations of the systems. Most of all, the users should remain in control. They should be able to retain access to their own data, determine their own privacy and data-sharing settings, and be able to override the system—or turn it off entirely—whenever they wish. "We need to make sure that technology serves us, and not that we serve technology," Augusto said.

As technology insinuates itself more deeply into our homes, we'll need more than ethical frameworks—we'll also need real legislation, with teeth, that protects consumers and their data, and a willingness to rein in developers that step over the line. In the last few years, some U.S. states have passed data privacy laws modeled on the EU's General Data Protection Regulation, and momentum for more sweeping federal regulation is building. But even the laws that do exist don't go far enough, and they haven't kept up with the pace of innovation. "They are usually running behind what the people creating the technology are doing," Augusto said.

The technology is pervasive and powerful, and buildings of the future won't simply harvest our physiological data, they'll *respond* to it in real time, becoming not just diagnosticians but caretakers. "You could have a building that looks after your well-being in an active way," said Holger Schnädelbach, an architect and researcher who specializes in "adaptive architecture." About a decade ago, Schnädelbach began work on a project he called ExoBuilding, a small, tentlike structure that responds to occupants' breathing. When the tent dweller inhales, the fabric walls balloon outward, as though the tent itself were a lung filling with air. When the person exhales, the walls contract.

"Initially, it was just a blue-sky bit of research," Schnädelbach said. "Pure joy in experimenting, really." But when he ran a tiny pilot study with a few of his university colleagues as guinea pigs, he discovered that people had surprisingly strong reactions to ExoBuilding. Sitting inside it was deeply relaxing, the test subjects said, and there was something hypnotic about watching the walls around them expand and contract with their own breathing. "In a matter of minutes, it had this weird bodily connection with people," Schnädelbach told me when I visited his lab at the University of Nottingham in late 2017. One volunteer even remarked that when the system abruptly switched off, he felt his own chest jolt.

That's not all. "We very quickly found that people quite substantially change the way they breathe in this thing, which was quite surprising," he said. Their respiration rates slowed and became more regular. ExoBuilding was providing an architectural form of biofeedback, making participants more aware of their breathing and helping them learn to control it.

This discovery made Schnädelbach realize that what started as a playful project might have real applications. In the years that followed, he installed versions of ExoBuilding, which he refined and renamed Breathing Space, in yoga studios and senior care facilities. In the future, it could be deployed in office spaces as a retreat for harried employees, Schnädelbach suggested.

Other researchers are conjuring up adaptive buildings that are far more complex. In a 2016 paper, scientists at the University of Castilla–La Mancha in Spain outlined a concept for an intelligent home that could manipulate our moods. Their system would track occupants' physiological reactions, body movements, and facial expressions, and then tailor the lights and music accordingly. "The ultimate aim is to maintain a healthy emotional state," the researchers write. If you're nervous or jittery, the house might turn on calming music and lights; if you're sad, it could cue cheerful choices instead.

The prospect of using architecture to manipulate people's physiological functioning and moods is obviously an ethically fraught one. "There's always dystopian and utopian versions of visions like that," Schnädelbach said.

So as the field moves forward, we as consumers need to speak up about the future that we want, demanding regulatory oversight and voting with our wallets. Though we can never eliminate all the risks, some of these products have enormous potential; deployed responsibly, they could yield real medical success stories. "It's something we still have an opportunity to get right," Augusto told me.

As we hurtle into the future, architecture provides one way that we can begin to take control of our own destinies. And there's more than one way for a home to save lives: our buildings will need to evolve not only to take advantage of new technologies but also to help us weather some of the global existential threats that are looming on the horizon.

EIGHT

HOPE FLOATS

L ATE ON A THURSDAY NIGHT in June 2017, what seemed like a simple summer thunderstorm rolled through southern Ontario. The water came fast and hard; in just a few hours, parts of the low-lying region saw more than a month's worth of rain. By morning, rivers had breached their banks, turning basements into swimming pools and running tracks into lakes. Local governments declared a state of emergency. Residents told local news crews that they couldn't remember ever seeing so much water in their lifetimes—some called it the kind of flood that comes along just once every century.

It was a fitting meteorological coincidence. Just a few days later, as the region was still wringing itself out, architects, engineers, and policy makers from around the world arrived in Waterloo, Ontario, to discuss how humans might adapt to our soggy, waterlogged future, one in which major floods become routine. They had one specific survival strategy in mind: amphibious architecture. Unlike traditional buildings, amphibious structures are not static; they respond to floods like ships to a rising tide, floating on the water's surface. As one attendee explained, "You can think of these buildings as little animals that have their feet wet and can then lift themselves up as needed."

The traditional approach to flood risk reduction has been to try to subdue the water—by erecting walls and levees and dams that hold it back. Amphibious architecture represents a different philosophy: its proponents believe that we can no longer fight against water; instead we have to learn to live with it. "The amphibious approach is an approach that suggests that *we* are the ones who do the adaptation," said Elizabeth English, an architect and engineer at the University of Waterloo, as she kicked off the second International Conference on Amphibious Architecture, Design and Engineering (ICAADE). English, who has fine features and a silver pixie cut, stood in front of a whiteboard trumpeting one of the conference hashtags: #floatwhenitfloods.

Flooding is the world's most common natural hazard; between 1995 and 2015, floods affected more than 2.3 billion people around the globe. Rising water will take an even heavier toll in the decades to come. Although the precise effects of climate change will vary widely depending on location, studies suggest we'll see heavier rainstorms, more intense hurricanes, and more frequent flooding. We'll also have more crop-withering droughts and heart-stopping heat. The wildfire season will grow longer and more destructive.

This isn't a distant possibility or some far-off apocalyptic future. It's our new normal. In just the few months following the Ontario floods, Hurricane Harvey dumped as much as sixty inches of rain over Texas, forcing tens of thousands of people from their homes; Hurricane Irma marched through the Caribbean, skipping from island to island like a deadly stone; and Hurricane Maria absolutely devastated Puerto Rico. Monsoons killed more than 1,000 people and damaged more than 800,000 homes in India, Nepal, and Bangladesh. Heavy rains in Sierra Leone triggered flash floods and mudslides that claimed some 600 lives, while flooding in Nigeria affected more than 100,000 residents. Record-breaking heat waves scorched parts of China and southern Europe, and fires consumed California, torching a quarter of a million acres.

Welcome to the Days of Drought. The Age of Wildfire. The Generation of Torrential Rain. Beyond the obvious immediate effects—loss of life and destruction of infrastructure—these extreme events can utterly destabilize communities and societies, destroying livelihoods, disrupting agriculture, triggering mass migration, and leaving deep psychological scars. (As many as 40 percent of disaster victims develop post-traumatic stress disorder.)

Climate change is an urgent, existential problem that requires sweeping, large-scale solutions, including a rapid shift away from fossil fuels. But even if we stop emitting carbon today (spoiler alert: we won't), we'll also need to adapt to the new world that we've already created. That means, in part, designing structures that are more resilient, able to withstand whatever gauntlet nature throws down. Resilient buildings can save lives and alleviate human suffering, mitigating the worst effects of the disasters looming on the horizon and helping us bounce back faster. And by treading lightly on our overcrowded planet and making the most of our dwindling resources, the best of these buildings may help stave off future catastrophes, safeguarding our survival in a turbulent future.

ELIZABETH ENGLISH HAS dedicated her career to understanding some of nature's most destructive forces. After earning a bachelor's degree in architecture and urban planning at Princeton, she went to MIT to study civil engineering. She wrote her thesis on wind, using the campus wind tunnel to study how blasts of air affected different types of buildings. In 1999, she moved to Louisiana. She spent a few years living in New Orleans and teaching at Tulane University before moving to Louisiana State University's Hurricane Center to conduct research on the trajectories of windborne debris.

In 2005, not long after she began her work there, Hurricane Katrina slammed into the Gulf Coast. The storm's high-speed winds peeled roofs off buildings and flung debris through windows, but it was the flooding that really shocked English. "Katrina was much more of a water event than a wind event," she said. Many New Orleans neighborhoods sit below sea level, protected by levees and flood walls. But as the hurricane churned over the city, these defenses failed catastrophically. Eighty percent of the city flooded, with some neighborhoods submerged in as much as twenty feet of water. Though the precise number of casualties remains uncertain, the storm is estimated to have killed nearly 2,000 people, most of them in Louisiana. It damaged more than 70 percent of the housing units in New Orleans and forced nearly all of the city's 450,000 residents to evacuate. Many never came back.

Those who did return faced an enormous rebuilding challenge. Because New Orleans remained vulnerable to future storms, the federal government recommended that residents permanently elevate their houses on raised foundations or stilts. That approach didn't sit well with English. "I started looking at the implications of all the flood damage and the social disruption that it caused, and I became very, very angry about the cultural insensitivity of the solutions that were being proposed," she told me.

English worried that hoisting the city's low-slung, shotgun-style houses into the air would ruin its sense of community, making it more difficult for residents to chat with neighbors and passersby. Elevation would require residents to navigate a flight of stairs whenever they went into or out of their homes, which would pose a particular challenge for the elderly and people with mobility problems. "People didn't want to move up," English said. "And it visually thoroughly destroyed the neighborhoods. There had to be a better way."

She discovered that better way in the Netherlands, where developers were building a cluster of amphibious homes in a flood-prone region along the Maas River. The concept wasn't completely novel; a handful of communities around the world, including the residents of Belén, Peru—an impoverished area sometimes called the "Venice of Latin America"—have devised amphibious shelters that rise off their foundations when it floods. But in the early 2000s, a team of Dutch designers and developers set out to make the approach more mainstream. The houses they built sat on hollow concrete boxes attached to large steel pillars. During a flood, the boxes would function like the hull of a ship, providing buoyancy. As the waters rose, the buildings would rise, too, sliding up the pillars and floating on the water's surface. When the waters receded, the houses would descend to their original positions.

It was an elegant solution, English thought, but not quite what she was looking for. Building a hollow foundation is a major construction project, and English wanted to give New Orleanians a cheap and easy way to modify their existing homes. In 2006, she founded a nonprofit called the Buoyant Foundation Project and started working with a group of architecture and engineering students to devise a method for retrofitting local homes with amphibious foundations.

One of the students told English about a remote Louisiana community on the banks of Raccourci Old River, a twelve-mile-long oxbow lake that connects to the Mississippi River. The area, where the student's family had a house, often floods in the spring, when the Mississippi surges. Several decades ago, some of the local residents had jury-rigged their own amphibious homes.

English's student took her out to the area during a spring flood. They toured the community by boat and saw dozens of homes—as well as the local restaurant and bait shop—floating, intact and undamaged, on the surface of the floodwaters. "It was brilliant," English recalled. "South Louisiana ingenuity."

English had invited several local residents, including David "Buddy" Blalock, to ICAADE to describe the system they'd devised, but shortly before the conference, Blalock had called to say that Raccourci Old River had flooded again. As English told us, "He called me and said, 'You know, I can't come. I'm eight feet off the ground and the water's not supposed to go down for another two weeks.'"

Blalock ended up dialing into the conference via video call, and as he sat inside his floating house, he shared his story. When he first moved to Raccourci Old River in 1983, he'd lived in a small trailer. But every time there was a flood, he had to move, pulling his trailer across the nearby levee until the water receded. The floods were less frequent in those years, but still, the back-and-forth wore on him. "I got tired of that," Blalock told us. "And I thought I'd look at other options. I'd done a lot of sailing, boating, and I thought, 'Why don't I build me a boat and live on a boat?' And so that's what I did."

Blalock built himself a home that essentially functioned as a pontoon raft, raising the structure a few feet into the air and attaching foam buoyancy blocks to its underside. He sank metal poles into the soil around his home, slid a hollow metal sleeve over each pole, and then welded each sleeve to the frame of the house. Each spring, when the floodwaters rose, the sleeves—and the house—would slide up the poles and then slide back down as the water ebbed.

Most residents evacuate the area when the flooding begins, but Blalock likes to stay put, waiting out the season in his floating house. "I've been out here riding the floods and have had as much as twelve foot of water under me," he said. "I've had waves crash into my porch," he added. "The waves can really build up to three feet or four feet and they rock it pretty good. But it doesn't creak and it doesn't move. The doors don't jam and the cabinets don't open." The house is so stable in floods that he's never had so much as a single piece of glass break.

When English first saw Buddy's house, she knew she'd finally found what she'd been looking for. A typical New Orleans shotgun house sits slightly above the ground, resting atop short piers; English thought that she could make it amphibious by fastening a steel frame to the underside of the house and affixing a set of foam buoyancy blocks to the frame. Rather than welding sleeves to the house, she'd attach the metal frame to telescoping guidance posts buried near the corners of the building. During a flood, the buoyancy blocks would keep the house afloat and the guidance poles would extend, allowing the house to rise up off its piers without floating down the street. (As the house rose, long, coiled water and electric supply lines would unfurl, while self-sealing breakaway valves would automatically close off the sewer and gas connections.)

English and her students built a preliminary prototype of the system, and in the summer of 2007 they put it to the test. Borrowing some corral panels from the College of Agriculture, they built a temporary flood tank around their model home and pumped in water straight from the Mississippi River. The tank filled with two, then three, then four feet of water. Slowly, the house began to rise. "It was a religious experience when it lifted off," English recalled. By the time they stopped pumping, the house was hovering about a foot above the piers.

The system can be installed by two reasonably handy people without heavy equipment for between ten and forty dollars a square foot, English told me. It's a bottom-up solution that puts homeowners in control and doesn't depend on major government investment. It leaves a building's appearance and structure almost unchanged and is both cheaper and more resilient than permanent elevation, which makes a building more susceptible to wind damage. And it's dynamic, adjusting to varying levels of flooding. "This is not a one-size-fits-all solution," English said, noting that the system would not extreme weather protection against high-speed water or waves. "But it's an excellent solution for some circumstances."

The approach works with, rather than trying to halt, natural cycles of flooding. Though they can be catastrophic for humans, periodic floods can have some ecological advantages, including recharging aquifers, redistributing sediment, and replenishing the nutrients in soil. The technique also completely reimagines water. "With amphibious construction, water becomes your friend," English said, noting that it's the water that lifts people to safety. "The water gets to do what the water wants to do. It's not a confrontation with Mother Nature, it's an acceptance of Mother Nature . . . If you pick a fight with Mother Nature, eventually you're going to lose."

English became so enamored of the approach that she began to think beyond the bayou—and about bringing amphibious architecture to some of the world's most vulnerable people. The hazards of climate change are not equally distributed; poor, disadvantaged, and marginalized communities will bear the brunt of the burden. These populations tend to live on the least desirable land, and a process of "climate change gentrification" is ensuring that that continues. In Miami, one of the American cities that is most threatened by sea level rise, homes located on higher ground are becoming more valuable; as the prices of these properties rise, low-income residents will be increasingly pushed into the more dangerous, low-lying neighborhoods.

Many people living in poverty can't afford to disaster-proof their homes or evacuate when a storm hits. And they often struggle to get back on their feet in the aftermath of disasters. Poor families are less likely to have insurance and tend to have fewer, and less diversified, assets to draw upon to rebuild their lives. They have less political power and may have less access to community and government resources. As a result of all these factors, climate change, and the extreme weather associated with it, has set off a vicious cycle that is exacerbating inequality.*

Hurricane Katrina laid many of these disparities bare. Many of the city's poorest residents were unable to evacuate because they did

---

*Climate change has already made poor countries poorer, two Stanford University scientists concluded in a 2019 paper.

not have cars—or a safe place to evacuate to. After the storm, many low-income households saw their home loan applications rejected, and the city's wealthy neighborhoods recovered faster than poor ones.

Buoyant foundations, English thought, could help vulnerable people ride out floods with less damage and rebuild their lives more quickly when the waters receded. In some cases, they might help keep communities intact—and even prevent people from being washed out of their ancestral homes.

In the years after developing her first prototype, she worked with the residents of Isle de Jean Charles, a thin slip of land that sits just southwest of New Orleans. For the last two centuries, members of the Biloxi-Chitimacha-Choctaw tribe have lived on the island, growing corn and wheat and pulling fish and oysters out of the bayou.

But the island is literally disappearing under their feet; thanks to a combination of rising seas and sinking land, Isle de Jean Charles has shrunk by more than 98 percent since 1955. Frequent floods and storms have forced many off the island. Tribal leaders have been hoping to relocate the entire community and reunite its scattered members, but the process has been slow, complicated, and politically fraught.

By the time English began consulting with residents in 2010, there were only twenty-six homes left on the island. Nineteen of them had already been put up on stilts; English and her colleagues drew up plans to amphibiate the rest. But before English could carry through on these plans, the government awarded the community $48 million to relocate. "I realized that they were in a very, very bad situation and amphibiating was just going to be a Band-Aid," English told me. "I didn't want to be a distraction in their moving forward in doing what they needed to do."

After moving from LSU to the University of Waterloo, she started working with First Nations communities whose reserves are threatened by floods. In addition, she and her colleagues at the Buoyant Foundation Project have sketched out amphibious-housing prototypes for low-income, flood-threatened regions in Nicaragua and Jamaica, adapting their concepts to use sustainable materials and meet local needs. Casa Anfibia, the Nicaragua design, calls for houses made of locally grown bamboo, which has a low carbon footprint, and swaps out the foam buoyancy blocks for recycled plastic barrels. It features a large wraparound deck, which residents can use to keep their pigs and chickens safe from rising floodwaters.

In 2018, the Buoyant Foundation Project retrofitted a handful of homes in Vietnam's Mekong Delta. (In a short video about the project, English leads her local collaborators in a rousing cheer: "Two, four, six, eight, we should all amphibiate!") The houses "performed beautifully" during the subsequent monsoon season, English told me, and the team is currently looking for funding so that they can scale up the project.

The goal, whether in Louisiana or Vietnam, isn't to build entire amphibious developments but to demonstrate that the approach is viable and then give homeowners the resources they need to do the retrofits themselves. "We go in, we show them that it's a possibility, we show them that it's cheap and easy and they can do it themselves," English said.

Amphibious buildings are popping up everywhere, from the United Kingdom, where the firm Baca Architects recently built a buoyant three-bedroom home on an island in the Thames, to Bangladesh, where one of English's students designed a sustainable, affordable housing unit that relies on eight thousand empty plastic water bottles for buoyancy. One ICAADE attendee presented a concept for amphibious health clinics in Thailand; another announced that he'd developed a new type of porous concrete that would be well suited for buoyant foundations.

(Some designers and engineers are proposing structures that are less amphibious than downright aquatic, designed to float permanently on water. Houseboats and floating villages have long existed in watery locales around the world—from the canals of Amsterdam to the Tonle Sap Lake in Cambodia—but these designs are something altogether different. At ICAADE, there was one proposal for floating high-rises, in which the rise and fall of the towers would produce electricity, and another suggestion for a floating Olympics, with all the events held in venues situated on the open ocean. The Dutch architecture firm Waterstudio has designed floating homes, restaurants, and hotels, as well as a floating golf course, spa, and mosque. In 2019, the company Oceanix unveiled its plan to build an entirely self-sufficient ten-thousand-resident 185-acre floating city.)

Still, amphibious structures remain more of a curiosity than a bona fide building trend, perhaps because the premise isn't exactly intuitive; when English first started telling people about her idea, they often laughed at her. The biggest hurdles, however, are more prosaic. In the United States, federal law requires most homeowners living in high-risk flood zones to purchase flood insurance, but buildings with amphibious foundations are not eligible for the subsidized policies offered by the National Flood Insurance Program (NFIP). In 2007, an official at the Federal Emergency Management Agency (FEMA), which administers the NFIP, urged English not to release more information about buoyant foundations and suggested that communities that permitted them could "jeopardize" their "good standing" with the NFIP.

When I reached out to FEMA in 2017, a spokesperson told me that the agency believed that more research was needed. "Although amphibious building technology is changing, these systems raise several engineering, floodplain management, economic, and emergency management concerns," the spokesperson said. "A technology that relies on mechanical processes to provide flood protection is not equivalent to the same level of safe protection provided by permanent elevation."

Caution is prudent, but the changing climate is making innovative approaches more urgent. "We're gonna be doing a lot more of this," Buddy Blalock told English back in 2017 as they paddled through one of Old River's frequent springtime floods. "There aren't many people out here that don't believe in climate change," he added.

English isn't giving up. She received a grant from Canada's National Research Council to test some new prototypes and develop guidelines for amphibious construction. And she can feel opinions about amphibious architecture beginning to shift. "I'll have people coming up to me and saying, 'I've never heard of this before, this is such a great idea, how can I do this in my community?'" English said. "People don't laugh at me anymore."

OF COURSE, we can't design our way out of climate change. Creating resilient neighborhoods, communities, and societies will require improving risk mapping, prediction, and early warning systems; developing sophisticated evacuation plans; preserving green space; investing in communication systems and emergency services; expanding insurance coverage; fostering diverse local economies; strengthening social ties; shoring up the social safety net; and, ultimately, tackling the root causes of poverty. And there are some places that we may need to simply abandon. But a resilient building *can* be the difference between a major catastrophe and a minor one, between a disaster and an inconvenience.

What that resilience looks like varies by location and depends enormously on geography and climate. Residents of Los Angeles or Oklahoma City have to guard against different dangers than denizens of New Orleans, so architects are thinking big about structures that can withstand all sorts of different disasters. In Joplin, Missouri, Q4 Architects designed a tornado-proof home by wrapping a perimeter of conventional rooms around an indestructible core that can be used as a refuge until the funnel spins past. Deltec Homes, a North Carolina–based company, makes wheel-shaped houses with aerodynamic features that are designed to resist hurricane-force winds. In Japan, which is known as a leader in seismic design, even skyscrapers have been engineered to withstand earthquakes, thanks to shock absorbers, vibration dampening pendulums, and Teflon-coated bearings.

Resilience needn't be radical. Hospitals in flood-prone regions, including New Orleans and Boston, are safeguarding against future storms by rethinking their layouts, putting administrative offices on the lower floors and moving the most critical areas, such as patient rooms and mechanical systems, to higher ones. In addition, we could save countless lives by updating our building codes, many of which are based on historical climate patterns that are rapidly becoming outdated, and by enforcing those that are already on the books. (After a major earthquake killed hundreds of people in Mexico City in 2017, a group of investigative journalists discovered that some developers had been flagrantly violating local building regulations.)

Although resilience is highly context-specific, it often goes hand in hand with sustainability. Natural disasters can knock out power and compromise water supplies, and architects are increasingly prioritizing what's known as "passive survivability," creating buildings that remain habitable even in such extreme circumstances. These buildings generally have lower energy demands—because they are well insulated, for instance, or rely on natural light and ventilation—and harness alternative energy sources, such as solar power.

And resilient design is about more than playing defense; sustainable, eco-friendly buildings can not only help us survive in a world transformed by climate change but also mitigate its worst effects. The typical modern building is an enormous resource hog. To create a single new building, raw materials have to be extracted and refined, parts manufactured and transported, structures assembled and secured. Building just one new single-family home creates about four pounds of waste per square foot, and the typical American home has more than doubled in size since the 1950s. Operating and maintaining buildings generates still more waste and requires enormous quantities of water and energy. The construction and operation of buildings is responsible for more than a third of all energy use worldwide and nearly 40 percent of energy-related carbon dioxide emissions.

Tackling climate change will require reducing the ecological footprints of our buildings, and architects, developers, and homeowners have grown more interested in creating structures that tread lightly on the land. Green building is a thriving industry, and sustainable rating systems and certification programs—from LEED to the Living

Building Challenge—have rapidly proliferated. Governments are doling out financial incentives to developers who embrace eco-friendly practices, and some consumers are deliberately downsizing their lives, moving into tiny homes and reconfigurable "micro-apartments."

With really careful design, we can even engineer buildings that give back, like "energy-positive" structures that produce more energy than they consume. Seattle's Bullitt Center, which has been called "the world's greenest office building," does just that, thanks to 575 solar panels arrayed on its roof. It also funnels rainwater into an enormous cistern in the basement; the water is filtered and treated so it can be used to meet the building's water needs. The wastewater from the center's sinks and showers is cleaned and pumped back into the ground, replenishing the local aquifer. The toilets send sewage to the building's basement, where it is transformed into compost.

Powerhouse, a consortium of mostly Scandinavian companies, is designing a portfolio of energy-positive schools, offices, and hotels, while Google's sister company Sidewalk Labs aims to create an entire "climate positive community" in the high-tech smart neighborhood it's developing in Toronto. To do so, it plans to deploy an array of green technologies, including a state-of-the-art electrical grid that incorporates solar and geothermal energy, digital stormwater management, smart trash chutes, and delivery and waste collection robots.

These splashy projects tend to get a lot of press, but resilience doesn't require a big budget or advanced technology. Hundreds of years before anyone had even conceived of LEED or smart grids, humans were building sustainable structures that made clever use of natural resources and were perfectly suited to their environments. Societies that settled in frozen climes used thick layers of snow and sod to keep their homes warm in winter. Desert communities constructed tall, densely packed buildings, which shaded residents from the blazing sun. People who lived in tropical locales boosted air circulation and captured cooling breezes by building elevated homes made of thatched palm and coconut leaves. Some inhabitants of hurricane-prone regions erected round huts with cone-shaped roofs that helped deflect high-speed winds; some in seismically active zones inhabited cave dwellings able to withstand earthquakes. These traditional building practices, based on centuries of hard-earned cultural knowledge, could help us create a world in which a safe, sustainable home isn't a luxury but a human right.

A VISION OF this very future, in which all people have access to affordable, resilient homes, sits on a small, sandy lot northeast of Los Angeles, in the high Mojave Desert. Early one Saturday morning in March 2018, I set out to see it, winding my way through the San Gabriel Mountains. As I descend, the baked brown landscape of Hesperia, California, comes into view. I take Maple Avenue, which has no maple trees to speak of, to Willow Street, which has no willows, past Hesperia High School ("Home of the Scorpions"), to Live Oak Lane—which has, you guessed it—until I reach a small housing development, neat rows of two-story homes in shades of tan. Just on the other side of the development, the road ends. There, amid sand and desert scrubs, small earthen domes dot the desert landscape, rising from the ground like strange adobe igloos.

This is the otherworldly architecture of the CalEarth Institute, a disaster relief charity and educational nonprofit that teaches people how to build resilient, sustainable homes out of the literal ground beneath their feet. Though it's an unseasonably cold morning, with gusting winds and storm clouds threatening overhead, more than a hundred people, bundled up in coats and hats, have driven out to this little patch of desert to honor CalEarth's founder, Nader Khalili, on the tenth anniversary of his death.

Khalili was born in 1936 in Tehran, one of nine children in a poor family. He studied Persian literature and architecture before deciding to try his luck in America. He arrived in San Francisco in 1960 with a backpack, a Persian-English dictionary, and $65. There, he worked a series of odd jobs—after mistranslating "stock market" as a place where farm animals were bought and sold, he turned up for a financial district temp job wearing his dirtiest work clothes—before eventually training as a draftsman and becoming a licensed architect.

He established practices in both Los Angeles and Tehran and spent years traveling back and forth between the two cities, working on urban, big-budget projects, like high-rise apartment buildings and hulking parking garages. And then one day, in what has become an oft-told family legend, he took his three-year-old son Dastan to the park. There, a group of local kids were racing one another down a tree-lined footpath. Khalili's son, the youngest and the smallest of the kids, competed enthusiastically but kept finishing in last place. "My little boy, getting to the starting point later than everybody else, comes to me at the fourth round, panting and with tears in his eyes, and says, 'Baba, Baba, I want to race alone,'" Khalili recalled in his memoir.

Khalili encouraged his son to run his own race. "This time he comes around, even later than usual, but he is happy and gives me a yellow leaf he has found on the way," Khalili wrote. "He has enjoyed his race, has had enough time to find a leaf, and above all he has come out first . . . Yes, I think to myself, it is a joy to race alone."

As he watched his son, Khalili thought about his own life choices. He was tired of ballooning budgets, of towering towers, of skylines ruled by steel and glass. He wanted to do something humble, to use natural materials to create homes for the people who needed them most. "What he really wanted to do was to find solutions for shelter," said Khalili's daughter, Sheefteh.

Having a safe place to bed down at night is a basic human right. "The right to life cannot be separated from the right to a secure place to live," a UN official wrote in a 2016 report. And yet, more than a billion people worldwide live in substandard housing, while 100 million are homeless.

In the mid-1970s, at the time Khalili was rethinking his career, these figures were lower but substantial, with hundreds of millions of people on the planet in need of safe, secure homes. Khalili wanted to provide them. In 1975, he closed his practices, got on his motorcycle, and rode off into the Iranian desert. For thousands of years, local villagers had built structures out of earth, fashioning mud and clay into sun-dried bricks. Khalili was attracted to the possibility of building with earth, and he began to brainstorm new ways to turn earth into sturdy, affordable homes.

Khalili, who returned to America in 1980, developed a system that he would come to call SuperAdobe, his own spin on a practice known as earthbag construction. He filled sandbags with moistened soil and arranged the bags in a large circle on the ground. He laid down a layer of barbed wire on top of the sandbags and then another slightly smaller ring of soil-filled sandbags on top. He continued in this manner, stacking alternating layers of sandbags and barbed wire atop one another until he formed a complete dome. The barbed wire essentially functioned like Velcro, sinking its small teeth into the sandbags and holding them in place. The whole dome could then be plastered over to make the structure permanent.

The dome wasn't an incidental part of the design—it was a critical one. Domes, which distribute forces more evenly than traditional, box-shaped houses, are innately strong. They're also aerodynamic and compact, with bases wider than their tops and low centers of gravity, making them less likely to sway or topple over. In the event of an earthquake, Khalili imagined one of his domes responding like a bowl turned upside down on a table. The bowl might slip and slide as the table shook, but it would not, he hoped, collapse. "The foundations are buried in the ground, but the houses will sit separate and free on top and let the earth shake in and out from underneath," he wrote.*

Khalili played with these concepts throughout the 1980s. In 1991,

*Khalili was not the first to discover the benefits of the dome. A number of Indigenous peoples have long built dome-shaped shelters, and in the 1940s the architect Wallace Neff began using enormous balloons, sprayed with concrete, to build "bubble homes." In 1954, the architect and inventor Buckminster Fuller patented the geodesic dome, a sphere made up of small repeated triangles. Though Fuller has passionate fans, the domes remain a niche building style, especially for private residences.

he bought seven sandy acres in the Mojave Desert and established the CalEarth Institute to act as his real-world workshop. Local building officials were dubious about his plans to construct buildings out of sandbags. "In fact, if we hadn't been trained to be courteous, we would have laughed out loud," two Hesperia planning officials wrote in an industry journal. But the SuperAdobe structures passed their seismic tests with flying colors. "[T]he required test limits were greatly exceeded," the officials wrote. The testing machinery failed before the buildings did.

Over the years, Khalili perfected his technique, which he patented in 1999. He discovered that the sandbags could be put in place first, while they were empty, and then filled, a breakthrough that eliminated the need to hoist heavy bags—and one that meant that even children could help build a SuperAdobe dome. "If you can carry one coffee can of soil, you're a full participant in this work," said Sheefteh Khalili, who took over CalEarth with her brother, Dastan, when their father died in 2008.

A small group of nonexperts can learn the method in a day, and then build a basic one-room dome on the following day, making Super-Adobe a good solution for disaster-relief housing, Sheefteh Khalili explained. The technique is sustainable both in materials and process, requiring little more than the earth under our feet—about as local as local gets—and our own human hands. The earth-filled sandbags make the structures both fire and flood resistant and provide insulation. (Domes, which have minimal surface area, relative to the amount of space they enclose, are also inherently energy efficient.)

SuperAdobe is versatile; it can be used to create domes of varying sizes, as well as vaults—long, narrow rooms with arched ceilings, like half a barrel on its side. Single domes and vaults can be connected to create large, multiroom homes. "And the cool thing is it's adaptable to the local culture and climate," said Sheefteh Khalili. The sandbags can be filled with nearly any kind of material—even volcanic ash or trash from a landfill can work.

SuperAdobe's first real-world deployment came after the first Persian Gulf War, when the United Nations Development Programme hired Nader Khalili to design fourteen shelters at a camp for Iraqi refugees. He trained UN staff members in his technique, and they subsequently trained the refugees, who built the shelters—each large enough for five people and requiring just $621 of materials—in less than two weeks. Khalili, a philosopher and poet at heart, saw a certain poetry in using SuperAdobe to shelter people who were fleeing political conflicts. "Sandbags and barbed wire, the materials of war, become the building blocks of peace and help refugees to resist war, as well as natural disasters," he wrote.*

During my visit to CalEarth, I wander around the institute's "emergency shelter villages," clusters of simple SuperAdobe domes that can be built quickly in the aftermath of a disaster. Stone paths wind between domes in a spectrum of earth tones. The shelters are humble—just one room with some small porthole-like windows and a Persian rug or two on the floor—but the thick walls feel sturdy, immovable, solid. Outside, the wind gusts and howls, but the domes are warm and quiet. They feel cozy, almost womb-like.

At the far end of one of the villages stands a small reddish dome that resembles an oversized anthill. This is Haiti One, which CalEarth staff designed after a catastrophic earthquake hit the Caribbean island in 2010. In the quake's aftermath, a CalEarth team flew to Haiti to interview some of the survivors, who were living in makeshift tent cities. "Anytime there's a refugee camp, they build it temporarily—it's tents and so on—but it ends up being there for years and years," Sheefteh Khalili told me.†

So she and her colleagues asked the displaced Haitians what kind of shelters they would need in order to feel safe over the longer term. "One is that it needs to be safe from a hurricane," she said. "A tent

---

*Khalili was especially taken with the work of Rumi, a thirteenth-century Persian poet. He published multiple books of Rumi translations and found inspiration in Rumi's line "earth turns to gold in the hands of the wise."

†She was right. Five years after the quake, nearly eighty thousand Haitians remained in what were supposed to be temporary camps.

would get washed away. Another was that they wanted to leave their children at home so they could go out and look for work, so it needed to have a locking door." Based on what they heard, the CalEarth staff designed a prototype for a new kind of post-emergency temporary housing: a ten-foot dome with three small nooks—one for sleeping, another for cooking, and a third for storage—and a locking door made from recycled shipping crates.

Although Nader Khalili's passion was providing homes for the poor, the homeless, and the displaced, he believed that SuperAdobe would be a good choice for anyone interested in creating a sustainable, disaster-resistant home, even the wealthy denizens of Southern California. "For many years people would come up from L.A. and be very, you know, eco, sustainable, conscious kind of people," Sheefteh Khalili told me. "And they'd be like, 'Yeah, this is all great, you know, but this is sort of beneath me. I wouldn't live in a little hut like this.'"

So Nader Khalili built Earth One, a three-bedroom, two-thousand-square-foot SuperAdobe home consisting of nine connected vaults. The home, which sits on the CalEarth campus, is fully furnished, down to the cozy woven blankets and the multicolored accent pillows. It's tricked out with modern amenities, including plumbing, central heating and air conditioning, a functioning fireplace, an open kitchen, and a two-car garage. It is, Sheefteh Khalili said, "kind of the American dream type of home, all built out of earth right there from the land." (Of course, if we truly want to become a sustainable society, we need to invest more heavily in public transit and design more walkable neighborhoods so that families don't need two cars.)

CalEarth hosts monthly open houses as well as weekend workshops and longer apprenticeships for those who are interested in building their own SuperAdobe structures. "This idea of empowering the individual is really at the core of what we do," Sheefteh Khalili said. "We show people that don't even know the right end of a shovel that within two days they can build a shelter." Some students eventually go on to teach their own classes and workshops, spreading the word about SuperAdobe.

There are now SuperAdobe structures in approximately fifty countries, including Venezuela, Madagascar, Australia, Hungary, Canada, Oman, India, Sierra Leone, and Japan. Some have already survived disasters. In 2015, a forty-dome SuperAdobe orphanage in Nepal withstood a 7.6 earthquake with just a few cracks in the external plaster, while nearby homes collapsed. Two SuperAdobe domes in Puerto Rico, built by an alumnus of a CalEarth workshop, reportedly survived Hurricane Maria completely unscathed.

In December 2017, the Thomas Fire, then the largest in California history, whipped through the grounds of the Ojai Foundation, an educational and environmental nonprofit in Southern California. The fire destroyed more than two dozen of the foundation's traditional buildings and shelters, but its SuperAdobe structure survived, even as the landscape surrounding it turned to ash. "For the people that saw that happen, I think that was really profound," Sheefteh Khalili said. "We've been able to show proof of concept, through the earthquakes, through the fires, to be able to tell people that we have a design that's fireproof, hurricane proof, floodproof."

Even so, I suspect that the appeal of SuperAdobe has its limits, and it's hard for me to imagine the area's wealthy enclaves truly embracing what are, essentially, homes made of dirt. After all, Americans' current housing preferences were on vivid display as I toured CalEarth—the McMansions at the neighboring housing developments literally loomed over the earthen domes. And the institute has faced setbacks; in 2017, the county informed CalEarth that it would not issue any more permits for SuperAdobe structures until the building method had been formally evaluated and approved by the International Code Council, which develops building safety standards that are widely used around the globe. That multiyear process is still under way, but if CalEarth wins approval, it should make it easier for people all over the world to earn permits for SuperAdobe buildings.

People are becoming more mindful of their ecological footprints and of the planet's limited resources, Sheefteh Khalili told me, and many are ready to try something new. In January 2018, at the first open house after the Thomas Fire, a large group of Ojai residents came to see whether CalEarth could offer them a better way to rebuild their neighborhoods. "The earthquake will be back, the tornado will be back, the fire will be back," Sheefteh Khalili said. "Can we be better prepared? Can we do something slightly different?"

Different doesn't necessarily mean SuperAdobe. There are many companies selling what they advertise as sustainable, resilient dome-shaped homes or dome-building kits at a variety of price points and in an array of materials. Japan Dome House sells "earthquake-resistant" domes made from a Styrofoam-like material; a Japanese resort installed more than four hundred of these domes, which were reportedly undamaged by a series of strong earthquakes in 2016.

Meanwhile, some American organizations have built villages of safe, sustainable "tiny homes" to house the homeless, and others are touting 3D printing as an answer to the affordable housing shortage. In 2018, ICON, a construction technology company based in Austin, Texas, used its "Vulcan" printer to create a 350-square-foot concrete house. The company—which said it ultimately aimed to print 600- to 800-square-foot structures in under twenty-four hours for $4,000 a pop—is turning its attention to Latin America, where it plans to print homes for impoverished families.

(The sustainability of 3D-printed buildings is a matter of debate. The technology can minimize waste and the need to transport construction materials across long distances, but some of the substances that the printers use, including concrete, are not especially eco-friendly. To fashion more sustainable structures, some designers are exploring loading 3D printers with greener materials, like bamboo fibers, sawdust, coffee grounds, and plant-based bioplastics.)

Many of these ideas are quixotic, but it's clear that we need to rethink how—and what—we build, perhaps in radical ways. The biggest housing problems we're facing—safety, sustainability, and affordability—are intertwined, and securing our future will require finding ways to tackle them all. "[My father] would always talk about how sustainability is this funny thing because people always talk about, 'Oh this is sustainable, this is LEED platinum certified, or this is such and such. It cost ten million dollars,'" Sheefteh Khalili told me. "And he would always say, 'You can't call that sustainable. It's not fair to do that because it's not affordable . . . People can't do it, they can't participate in it.' If it's going to be truly sustainable, it has to be accessible."

What Nader Khalili wanted was to devise a truly universal form of architecture. And when he said universal, he really meant it.

NINE

BLUEPRINTS FOR
THE RED PLANET

OVER THE COURSE of his career, Nader Khalili became an expert on how to make use of limited resources to build homes in some of the most inhospitable places on the planet. But this planet was just the beginning. Khalili's ideas, of making the most of what we found beneath our feet, were perfectly suited for building human habitats in outer space. As he once explained, "There are even fewer resources on the Moon than in our deserts, and the climate is much harsher."

In 1984, Khalili presented several of his design concepts at a NASA-sponsored symposium, "Lunar Bases and Space Activities of the

21st Century," in Washington, D.C. Before a crowd of scientists and engineers, he outlined how astronauts could build shelters out of lunar regolith, the layer of soil, rocks, and debris that covers the Moon's surface. This regolith is rich in crushed basalt, a volcanic rock that forms when lava cools. Lunar construction crews could pile this pulverized rock in a large mound and then use a giant mirror or magnifying glass to focus the sun's rays on its surface, Khalili suggested. The heat would melt the basalt on the surface, which would begin to flow down the mound and then harden as it cooled; colonists could then dig out the regolith underneath, leaving behind a sturdy dome. They could use a similar approach to create "lunar adobe" bricks, or even shape melted moonrocks on what he described as "a centrifugally gyrating platform—a giant potter's wheel."

In the years after the symposium, Khalili consulted with scientists at NASA, the Los Alamos National Laboratory, and McDonnell Douglas Space Systems as he honed his concepts. It was in those years that he developed SuperAdobe, which he thought would be well suited to the Moon. By stuffing regolith into flexible tubes or bags, lunar settlers could build themselves safe, durable shelters in harmony with alien landscapes. "Discovering suitable dimensions of blocks, techniques of construction, and appropriate material composites while developing their own sense of unity with the lunar entity can be the start of human independence from Mother Earth, creating shelters in the heavens," Khalili wrote.

While many architects are devoting themselves to resilience, and ensuring that our current homes remain habitable well into the distant future, others are turning their attention to new frontiers. What if we can't stay put—or don't want to? Where might we go next? A natural next step for humanity is to expand into space, to build outposts, settlements, cities, and societies on the Moon, Mars, and beyond.

"It's inconceivable to me that humanity will only ever exist on one planet," said Brent Sherwood, an architect and aerospace engineer who serves as the vice president for advanced development programs at Blue Origin, the private space company founded by Jeff Bezos. Sherwood was born in 1958, the same year that NASA was created, and he came of age during the high-flying Apollo era. He was taken with the romance and adventure of space exploration and was completely convinced, even as a boy, that humanity was destined to build settlements among the stars. "It was obvious at the time that we were going to be building cities on the Moon," he told me. And Sherwood wanted in on it. He carved out a career path that would set him up nicely for lunar urban planning, earning advanced degrees in both architecture and aerospace engineering.

After grad school, he went to work for Boeing, which later landed the NASA contract to build the International Space Station (ISS). He spent nearly two decades at the company, working on the ISS and other projects. In 2005, he moved over to NASA's Jet Propulsion Laboratory (JPL), where he spent years devising ideas for missions that would help scientists unravel our solar system's mysteries, before decamping for Blue Origin in 2019.

Throughout all his day jobs, he's never stopped dreaming of cities on the Moon. He's part of a small but passionate community of designers and engineers who are drawing up ideas for extraterrestrial settlements, mostly in their free time. "There's really very little professional work in it, so pretty much everyone in the field does something else to pay the mortgage," said Sherwood, who also serves as the chair of the Space Architecture Technical Committee of the American Institute of Aeronautics and Astronautics.

They're responding to a real interest in creating long-term human habitats in space. The United States, China, Japan, and Russia have all expressed a desire to build Moon bases, while the European Space Agency has called for the construction of an international "Moon village." Several Silicon Valley billionaires have their hearts set on colonizing space: Jeff Bezos wants to establish "a sustained human presence on the Moon," while Elon Musk is aiming for Mars. NASA hopes to send humans to the Red Planet in the 2030s, and the United Arab Emirates has announced plans to establish a Mars colony within the next century.

Permanent settlements on the Moon or Mars could serve as research bases, housing geologists, astronomers, and cosmologists who are trying to answer fundamental questions about the universe. They could be lucrative business opportunities for companies interested in mining space minerals or bringing adventurous tourists on the vacation of a lifetime. They could become way stations and staging grounds for voyaging even deeper into the cosmos and open up new opportunities for humanity.

And in the distant future, they might even act as an insurance policy, increasing the odds that our species endures for many millennia to come. Though our short-term survival depends on becoming better stewards of Earth—colonizing space is an extremely long-term project—we can't prevent every possible planetary catastrophe. We could be obliterated by an asteroid, and within the next several billion years, the expanding sun is likely to scorch Earth's surface, evaporate its oceans, and spur mass extinctions. If we want to survive, we'll need to have more than one home.

The irony is that our continued existence may hinge on figuring out how to live in environments that are literally lethal. For all the talk about their habitability, the Moon and Mars both offer endless opportunities to die. Temperature swings are extreme; the Moon can range from hundreds of degrees below zero at night to hundreds above during the day. (Mars can be equally cold at night, though daytime temps top out at a balmy 70 degrees Fahrenheit.) Unlike Earth, neither body is protected by a robust atmosphere or magnetic field; as a result, human settlers will be exposed to dangerous levels of radiation, from solar flares and cosmic rays. The lack of atmospheric pressure—and oxygen—means that colonists won't be able to set foot outside without wearing space suits.* The Moon is regularly bombarded by meteoroids, while enormous, weeks-long dust storms are not uncommon on Mars.

Such extreme environments will demand a lot of our buildings, and if we want to settle in space, we'll need to create structures that can withstand—and protect us from—all of these hazards. These dangers also mean that no matter how big our space colonies get, they will exist almost entirely indoors. "All architecture on the Moon or Mars is by definition an interior architecture," Sherwood told me. "Because the real exterior is lethal."

Again and again, in countless studies here on Earth, scientists have demonstrated the impact that our buildings can have on our health, behavior, and happiness. They shape our sleep patterns and stress levels; our diets and moods; and our physical fitness, job performance, immune responses, and social interactions.

In space—where settlers are likely to make spending 90 percent of the day indoors look like nothing—the effect of every design decision will be magnified. Space architecture presents an opportunity to apply everything we've learned about creating supportive indoor spaces— and to gain new insight into what we need in order to live healthy indoor lives.

Space architects are embracing the challenge. "Everybody I know in the field is emotionally committed, deeply viscerally committed, to the idea that it's appropriate and a natural future step for humans to be living in space," Sherwood told me. While these extreme environ-

---

*Some scientists dream of deliberately reengineering the Martian climate to make it more human-friendly, a process known as "terraforming." One approach would involve releasing greenhouse gases on the planet, including the carbon dioxide trapped in the Martian ice caps. These gases would form a thick atmosphere that would cover Mars like a blanket, gradually warming the planet. Terraformers could then plant trees and shrubs to generate oxygen. The end result, in theory, would be a warm, oxygenated planet that more closely resembles Earth. But these proposals are highly speculative and controversial, and even if they're technically feasible, they'd take centuries (at least). For the foreseeable future, what we see on Mars is what we get.

ments present unique difficulties, and will require some novel architectural solutions, the biggest challenge is more familiar: How can we make these remote outposts feel like home?

AS KHALILI HIMSELF INTUITED, the best way to build in space will be to make clever use of the resources we find there. In its simplest form, this might entail repurposing preexisting geological features. The Moon and Mars had turbulent volcanic pasts that have left behind underground caves and lava tubes, which could theoretically provide shelter from surface-level hazards.

But building safe, controlled environments inside these natural features is likely to be difficult. "Living in lava tubes is a sort of romantic but impractical idea," Sherwood told me. It may be easier to create our own structures from scratch, using some of the chemical compounds and minerals that *are* abundant on the Moon and Mars. "Mars is very rich in iron, which means we think we can smelt the iron from the ores and create very, very large structures," said Madhu Thangavelu, a space architect who lectures at the University of Southern California.

Or we could mix up batches of "Mooncrete" and "Marscrete" by mixing water with lunar or Martian regolith. (Though there's little liquid water on the Moon or Mars, we could extract it from ice or synthesize it from hydrogen and oxygen present in the regolith.) Or we could swap out the water for sulfur, which is plentiful in lunar and Martian soil. Simply extract the sulfur, heat it until it liquefies, mix it with a handful of moondust, and voilà: waterless concrete, perfect for all our Moon masonry needs. Engineers at Stanford University have also proposed using genetically engineered bacteria to produce proteins that can bind regolith together into a kind of "bio-concrete."

We could shape our space concrete into bricks or pour it into molds—or we could load it into 3D printers. In 2013, the European Space Agency unveiled a 3D-printed 1.5-ton block made of simulated lunar soil mixed with chemical binders. We could skip the chemical additives altogether by deploying solar-powered 3D printers that use focused beams of light to fuse thin layers of unadulterated regolith together into solid bricks, a technique known as "solar sintering." (The approach is, in some ways, a more modern version of Khalili's suggestion that we melt regolith with an enormous magnifying glass.) In 2017, ESA researchers used a similar process to create hard, rust-colored moon bricks in just a few hours.

The versatile technology of 3D printing holds enormous potential for space architecture, and there are a lot of ways to employ it. In 2015, NASA launched its 3D-Printed Habitat Challenge, a $3 million multiyear contest to solicit concepts for new kinds of space shelters. The team that won the first phase proposed printing dome-shaped ice houses—what would be, in essence, space igloos. Their design called for robots to harvest and heat the ice on Mars and then deposit this liquid water, along with an insulating gel and a fibrous additive, in thin, layered rings. In the frigid Martian climes, these layers would rapidly freeze, forming a strong, translucent structure. Another team proposed 3D printing a tall egg-shaped structure out of basalt fibers and renewable "bioplastic" derived from plants that astronauts could grow on Mars.

Designers are also drawing up plans for inflatable structures, which are relatively easy to transport and erect—we'd just have to pump them full of pressurized air once we arrived at our extraterrestrial destinations. NASA's been sketching out designs for inflatable space habitats since the 1960s; one of its most ambitious concepts was the TransHab, an inflatable barrel-shaped module intended to house astronauts working on the ISS. The multifloor habitat, which the agency designed in the late 1990s, would have been a roomy 12,000 cubic feet when fully expanded, with room for a kitchen, individual sleeping areas, exercise equipment, and a "Full Body Cleansing Compartment." It would have been protected by layers of insulating fabric, foam, and woven Kevlar. The project was abandoned after Congress eliminated its funding, but inflatables remain an area of active research and development. In 2016, astronauts successfully attached the inflatable Bigelow Expandable Activity Module (BEAM), developed by Bigelow Aerospace, to the ISS.

Other architects and engineers have proposed rocket ships and Mars landers that transform themselves into more permanent homes; folded, origami-like shelters; structures that self-assemble, thanks to magnetic tiles or smart, "shape memory" plastics; and spinning space stations that produce their own gravity. "It's possible that the right type of interior design is going to be something that people from Earth would find really odd," Sherwood said.

BUILDING SETTLEMENTS IN space would be a major technological achievement, but the accommodations will not be luxurious, especially in the early years. Much of the research so far has focused on creating what is essentially a minimum viable product—an efficient, economical shelter that would allow us to survive in environments that would otherwise quickly kill us. These structures will be isolated and austere, far from everything we know and love on Earth and missing many of its creature comforts. We could go days without leaving homes that are buried beneath layers of rocks and soil. Thangavelu told me that he found this profoundly ironic: "We would fly in as the overlords of the planet, and then guess what? Go live like cavemen for the rest of our lives."*

These living conditions would take a real toll on our well-being. Small, carefully screened crews of astronauts could probably tough it

---

*Not everyone takes this as a given. Sherwood believes that early space settlements will cater to rich adventurers. "The customers who can pay to enable the early opportunities are very high-end 'tourists,'" he told me. "Luxury in space certainly won't mean big rooms, but it probably will mean grand views, great linens and finishes, fantastic food, outstanding service—all the things that make high-end travel what it is on Earth today." And we may, in fact, see a range of space accommodations—luxe living quarters for space tourists as well as more utilitarian spaces for scientific and work crews.

out for a few months or even years. But if we truly want to become an interplanetary species, we'll need to create shelters that help us to not just survive but thrive in these strange new worlds.

While the external environment will be vastly different in space, *we* won't be, and even the most far-flung buildings will need to meet our basic human needs. Fortunately, we've learned a lot about how thoughtful design can help keep our bodies strong, our minds sharp, and our morale high. And we'll need to reach deep into our arsenal of evidence-based design strategies to create livable spaces in space. "I don't have to be a Martian to understand how to make [habitats on Mars]," said Vera Mulyani, the founder and CEO of Mars City Design, a California-based company dedicated to drawing up plans for sustainable cities in space. "I just have to be human."

One of the biggest lessons from our research here on Earth is how absolutely critical regular exposure to daylight is to our well-being. We can be fairly confident that spending day after day in a dark underground Moon bunker would wreak havoc on our moods, productivity, and health. Though radiation is a real danger, we'll need to find a way to bring at least some daylight into our lives. That's one of the reasons the idea of making habitats out of ice has such promise. As it happens, water (and water ice) absorbs some of the high-energy, short-wavelength radiation that poses a hazard to human health but not the slightly longer wavelengths that make up the visible spectrum. So a space shelter made of ice could conceivably protect its inhabitants from harmful radiation while letting the sun shine in.

There are Earthbound buildings we can draw inspiration from as well. Consider how you feel when you step inside a Gothic cathedral. Though they are made of stone and rarely provide expansive views of the outdoor landscape—they often feature stained-glass windows that diffuse and scatter the light—they don't feel at all oppressive. "It's the opposite," Sherwood said. "It's inspirational, and it's supposed to connect you with the ineffability of heaven so it's all about light at the top and these sort of soaring structures. Just because it's an interior architecture doesn't mean it has to be like living in a cave."

While we're considering light, we'll need to figure out a way to keep our circadian rhythms ticking away with some degree of normalcy. Our biological clocks evolved here on Earth and are closely pegged to its twenty-four-hour day.* They'll be profoundly thrown off by the day and night cycles of the Moon, which last roughly 28 days. That means that many locations on the Moon receive two weeks of constant sunlight followed by two weeks of unending darkness.†

The Mars day, or sol, is closer to our own, clocking in at 24 hours and 39 minutes (and 35 seconds, to be exact). It may not sound like a huge difference, but after eighteen days on Mars, you'd be nearly twelve hours off Earth time. Eighteen days later, you'd be almost back in sync. And on and on it would go. Over the long term, it's possible that our bodies would adapt, but if we don't, we could find ourselves battling never-ending jetlag. Either way, carefully calibrated lighting could help minimize the disruption. In 2016, NASA replaced the fluorescent bulbs on the ISS with programmable LEDs that put out energizing, bright blue light in the morning and a dim, sleep-friendly amber at night. The agency is studying whether this circadian lighting scheme improves astronauts' sleep and cognitive performance.

Space will sever our connection with nature in other ways, too. The landscapes on the Moon and Mars are bleak and barren; when we do gaze out the windows of our extraterrestrial homes, we won't see grassy fields or gently rustling willows. Even in the densest cities on Earth, we can go outside and breathe in fresh air or feel the wind, rain, and sun on our skin. In space, we won't even have that.

But we can try to bring in some design elements that keep space settlers tethered to Earth's ecosystems and landscapes. The Soviet

---

*Our intrinsic clock is *slightly* off the solar one—our bodies tend to run on cycles that last 24.18 hours, on average. But daily exposure to sunlight and other time cues keeps our clocks coupled to the twenty-four-hour day.
†There are places on the Moon—especially on mountains, in craters, and near the poles—that experience longer periods of daylight or darkness.

space program has long recognized the benefits that plants can have for astronauts' mental health. Their very first space station—Salyut 1, launched in 1971—included a small greenhouse named Oasis, and the cosmonauts spoke reverently about their botanical charges. "These are our pets," one reportedly said. His crewmate went further: "They are our love." Another cosmonaut apparently slept beside the greenhouse so he could see the plants first thing every morning.

The Soviets included greenhouses and gardens on many of their subsequent space stations and missions; the Russian side of the ISS contains a wall-mounted greenhouse. In 2003, after the space shuttle *Columbia* disintegrated, killing all aboard, Russian officials tried to calm the cosmonauts aboard the ISS by telling them to spend more time gardening. "It can be like meditation," said Sandra Häuplik-Meusburger, a space architect at the Vienna University of Technology who has interviewed astronauts and cosmonauts who have spent time on the ISS. "It allows the mind to wander off, and suddenly you feel like you have much more space than you have in reality."

We can help ensure that space settlers get a healthy dose of nature by piggybacking on our food production systems, providing views from our living spaces directly into our greenhouses. One of the teams in NASA's 3D-Printed Habitat Challenge—led by Vera Mulyani, as it happens—has proposed carpeting space greenhouses with moss, which astronauts could walk across to reawaken their tactile connection with nature. We could add nature photos and murals, pipe in nature sounds, and create virtual reality systems that enable astronauts to take walks in the simulated woods. It's not a perfect replacement for the real thing, but it can have some of the same psychological benefits.

These features could go a long way toward reducing some of the stress and boredom of what will essentially be house arrest. Though space settlers will presumably be volunteers, in the early years at least, their lives will be circumscribed in ways that would be recognizable to many of the people who are incarcerated on Earth. They'll live, sleep, socialize, and work in the same small indoor space, day in and day out. Their homes will have few frills and little in the way of pleasant sensory stimuli. Their environments and days will likely become routine and then monotonous.

Designers could provide some much-needed visual variety by strategically deploying different kinds of lights, colors, patterns, and materials. They could even incorporate art. In a 2011 paper, a pair of behavioral scientists proposed using lights to create artificial rainbows, building lunar Zen gardens, and erecting colorful solar panels in the shape of giant flowers.

Astronauts' social lives will be circumscribed, too. The first crews and settlers will be wrenched out of their familiar social networks and thrown into tight quarters with people who may be near strangers. They'll live and work alongside the same people, all day, every day, with no way to escape. "You cannot just break away and go for a walk and reconstitute your mind and come back," Thangavelu explained.

Space agencies invest a lot in trying to assemble compatible teams, but interpersonal tension is inevitable. As one cosmonaut once put it, "All the conditions necessary for murder are met if you shut two men in a cabin measuring 5 m by 6 m and leave them together for two months." Though there haven't been any murders in space, there have been brutal physical attacks in other isolated, confined environments. (A Soviet scientist sealed inside an Antarctic research base for the winter reportedly became infuriated by a game of chess and killed his opponent with an ax.)

Scientists have documented interpersonal problems in studies that are designed to replicate the conditions of living in space. Häuplik-Meusburger has observed volunteers sealed inside a simulated Mars habitat located on the side of a Hawaiian volcano. The habitat, a two-story geodesic dome, is part of the Hawai'i Space Exploration Analog and Simulation (HI-SEAS) project, which started running imitation Mars missions in 2013.

In August 2015, a crew of six adults—three women and three men—entered the habitat for a yearlong stay. During the mission, the crew members behaved as though they were actually living on Mars. They followed strict daily routines, prepared their own food, maintained their own equipment, and performed scientific experiments. They had to don spacesuits whenever they left the dome and contend with a forty-minute delay when they communicated with outsiders. Over the course of that year, they had no face-to-face encounters with anyone who lived outside the dome.

Tensions flared and the group quickly splintered. "Activities which involved the whole crew were often forced or otherwise avoided, due to social rifts which formed very early in the mission and were exacerbated by continued time living within the dome," Häuplik-Meusburger and her colleagues reported. Group dinners grew tense, and people started dropping out of the regular movie nights they'd initially arranged. The longer the mission went on, the more the crew members craved privacy, and some began to spend hours holed up in their own individual bedrooms. Even then, the lack of soundproofing made it difficult for crew members to truly escape. "Privacy is not just physical closure but also auditory and probably also olfactory privacy," Häuplik-Meusburger told me. The layout made matters worse; the bedrooms, dining room, kitchen, and exercise area were all easy to see from the large common area, which created, one participant reported, "a sense of observation which is difficult to escape from."

Even in small habitats, it will be crucial to give all space settlers at least one truly private space. Architects would also do well to take a page or two from the universal design playbook by providing spaces that allow for different levels of social interaction—some private and low-key, others more public and bustling. Beyond private bedrooms and bunks, we could give astronauts individual ownership over other small zones or areas, like garden plots to be used however they see fit.

As colonies and settlements grow, designers will need to construct spaces where larger groups of people can gather. "The architecture to support the community is different from the architecture to support a small crew of professionally trained astronauts," Sherwood told me.

"What's the equivalent of a town square or an auditorium in a place like the Moon?"

One idea we could borrow from Earth, Sherwood said, is the shopping mall. A mall is essentially an indoor, climate-controlled main street—with plazas and fountains and trees—where people can stroll and mingle. Sports facilities, game rooms, and movie theaters could provide spaces for play, an important part of keeping space settlers sane and happy.

Sports and games will be a new challenge on the Moon and Mars, however, which have much weaker gravitational fields than Earth does: the gravity on the Moon is just one-sixth as strong as it is on Earth; on Mars it's three-eighths as strong. This relatively weak gravity won't just change the trajectory of a soccer ball—it could also present real health risks. Earth's strong gravity may feel like a bummer, especially when you drop a glass or fall off a bike, but our constant fight against it is part of what helps keeps our bodies strong. When astronauts do long tours on the ISS, where they don't feel the tug of gravity at all, their bones and muscles begin to break down. "The only way you can be there for a few months at a time is to exercise like two hours a day," Sherwood said. Even then, it can take astronauts months to rebuild their strength after they return to Earth.

We won't be entirely weightless on the Moon and Mars, but it's possible that their weak gravity could do similar damage to our bodies, especially over the long term. If that turns out to be the case, incorporating active design strategies—like designing pleasant indoor walking paths and including lots of stairs—might help reduce the number of hours we have to spend sweating it out on our space treadmills.*

We'll also need to draw on all our knowledge about sustainability and resilience. Lunar and Martian bases will be extremely remote,

---

*Microgravity causes other problems that aren't so easily addressed with exercise. For instance, in the absence of a strong gravitational force, the fluids in our bodies shift around and flow upward into the chest and head. This redistribution of fluid can affect our cardiovascular systems and flatten our eyeballs, distorting our vision.

and they'll have to withstand failures, shocks, and disasters. Resources will be scarce, so space buildings will have to be supremely efficient; we'll have to recycle and reuse everything we possibly can. (Yes, including our own waste.)

Even if we never move to Mars, developing structures and systems that meet all of these requirements could pay dividends right here on Earth. NASA has already developed technologies, including increasingly powerful solar cells and wastewater recycling systems, that are boosting the performance of our everyday buildings. Likewise, some of the materials that engineers have developed for space construction—like sulfur- and bio-based concretes—can be more sustainable than our traditional building materials; deploying them on Earth could help reduce our carbon footprint and slow the pace of climate change.

In this way, plotting out what it would take for humanity to relocate to another planet in the event that this one becomes uninhabitable could help us avoid that very outcome. Planning space settlements can teach us how to live more thoughtfully and responsibly on our own ailing, overcrowded planet and how to build homes in hostile environments as conditions on Earth deteriorate. It could be, in its own way, a way of reversing our destiny—it's not precisely what Arakawa and Gins, the death-defying duo, had in mind, but it might just help ensure that our species continues to survive long into the future.

What's more, learning how to live comfortably in these extreme, extraterrestrial worlds could help us generate design ideas that we can appropriate for our homes here. In 2017, that possibility prompted IKEA, the Swedish retail giant, to send several of its designers to live in a simulated Mars habitat in Utah. There, they learned what it would be like to live on a cramped space station—and what it might take for astronauts to feel at ease. The experience inspired them to create a collection of compact, lightweight products, including air purifiers and terrariums, intended for Earthlings who happen to live in small apartments. (The team is also reportedly planning to release an air-purifying fabric.)

In space, we'll be starting with a blank slate. "This is the opportunity for the next generation to imagine what kind of life, what kind of home and environment they want," Mulyani told me. What kinds of buildings, settlements, and cities do we really want to build? What do we want them to look like, and how do we want them to function? What aspects of our lives on Earth do we want to bring with us, and what would we prefer to leave behind?

As we fantasize about these Moon- and Marstopias, we should ask ourselves how well what we've built here on this planet lives up to those ideals. It's not too late to change course. We have the tools and techniques to build a happier, healthier world, whether that's here on Earth's firm, familiar ground—or somewhere far beyond.

INTRODUCTION

3 *on a busy street*: Information on the Mitaka lofts is from: "Reversible Destiny Lofts MITAKA—In Memory of Helen Keller," Reversible Destiny Foundation, accessed April 25, 2019, www.reversibledestiny.org/architecture /reversible-destiny-lofts-mitaka. "Reversible Destiny Lofts MITAKA," accessed April 25, 2019, www.rdloftsmitaka.com/. Stephen Hepworth, interview by author, October 17, 2018. Miwako Tezuka, interview by author, October 17, 2018. S. T. Luk, interview by author, October 17, 2018.

4 *"old-fashioned"*: Shusaku Arakawa and Madeline H. Gins, *Mechanism of Meaning* (New York: Harry N. Abrams, 1979), preface.

4 *"immoral"*: Fred A. Bernstein, "A House Not for Mere Mortals," *The New York Times*, April 3, 2008, www.nytimes.com/2008/04/03/garden/03destiny .html.

4 *"That mortality has been"*: Madeline Gins and Shusaku Arakawa, *Architectural Body* (Tuscaloosa: University of Alabama Press, 2002), xv.

4 *"We believe that people"*: Gins and Arakawa, *Architectural Body*, xvi.

4 *In Yoro, Japan*: "Site of Reversible Destiny—YORO," Reversible Destiny Foundation, accessed April 25, 2019, www.reversibledestiny.org/architecture /site-of-reversible-destiny-yoro. Hepworth, interview, October 17, 2018. Luk, interview, October 17, 2018.

4 *In East Hampton, New York*: "Bioscleave House (Lifespan Extending Villa)," Reversible Destiny Foundation, accessed April 25, 2019, www .reversibledestiny.org/architecture/bioscleave-house-lifespan-extending -villa. Hepworth, interview, October 17, 2018. Luk, interview, October 17, 2018.

5 *In fact, their spaces*: Tezuka, interview, October 17, 2018.

5 *"cities without graveyards"*: The Solomon R. Guggenheim Foundation, *Reversible Destiny—Arakawa/Gins* (New York: Guggenheim Museum Publications, 1997), 239.

5 *Arakawa passed away . . . four years later*: Fred A. Bernstein, "Arakawa, Whose Art Tried to Halt Aging, Dies at 73," *The New York Times*, May 20, 2010, www.nytimes.com/2010/05/20/arts/design/20arakawa.html. Margalit Fox, "Madeline Arakawa Gins, Visionary Architect, Is Dead at 72," *The New York Times*, January 12, 2015, www.nytimes.com/2014/01/13/arts /design/madeline-arakawa-gins-visionary-architect-dies-at-72.html.

5 *Those who wish*: "Reversible Destiny Lofts/for2people," Airbnb, accessed April 25, 2019, www.airbnb.com/rooms/4606903?guests=1&adults=1.

5 *indoorsy*: With apologies to the comedian Jim Gaffigan, who first introduced me to the wonderful term "indoorsy."

6 *North Americans and Europeans*: Neil E. Klepeis et al., "The National Human Activity Pattern Survey (NHAPS): A Resource for Assessing Exposure to Environmental Pollutants," *Journal of Exposure Science and Environmental Epidemiology* 11 (2011): 231–52. World Health Organization, *Combined or Multiple Exposure to Health Stressors in Indoor Built Environments* (Copenhagen: WHO Regional Office for Europe, 2014).

6 *The island of Manhattan*: Laura J. Martin et al., "Evolution of the Indoor Biome," *Trends in Ecology and Evolution* 30 (2015): 223–32.

6 *Over the next forty years*: United Nations Environment Programme, *Towards a Zero-Emission, Efficient, and Resilient Buildings and Construction Sector: Global Status Report 2017* (2017), 13.

6 *Women who give birth*: Ariadne Labs and MASS Design Group, *The Impact of Design on Clinical Care in Childbirth* (2017), https://massdesigngroup.org

/work/research/impact-design-clinical-care-childbirth. Neel Shah, interview by author, December 21, 2016.

6 *Warm, dim lighting*: Nino Wessolowski et al., "The Effect of Variable Light on the Fidgetiness and Social Behavior of Pupils in School," *Journal of Environmental Psychology* 39 (2014): 101–108.

6 *Fresh, well-ventilated air*: Joseph G. Allen et al., "Associations of Cognitive Function Scores with Carbon Dioxide, Ventilation, and Volatile Organic Compound Exposures in Office Workers: A Controlled Exposure Study of Green and Conventional Office Environments," *Environmental Health Perspectives* 124 (2016): 805–12.

6 *In a 2016 study*: Ian R. Drennan et al., "Out-of-Hospital Cardiac Arrest in High-Rise Buildings: Delays to Patient Care and Effect on Survival," *CMAJ* 188 (2016): 413–19.

7 *In one study*: Sheldon Cohen et al., "Apartment Noise, Auditory Discrimination, and Reading Ability in Children," *Journal of Experimental Social Psychology* 9 (1973): 407–22.

7 *subsequent research has confirmed*: Brianna T. M. McMillan and Jenny R. Saffran, "Learning in Complex Environments: The Effects of Background Speech on Early Word Learning," *Child Development* 87 (2016): 1841–55. Kristine Grohne Riley and Karla K. McGregor, "Noise Hampers Children's Expressive Word Learning," *Language, Speech, and Hearing Services in Schools* 43 (2012): 325–37.

7 *Scientists have known*: There are many papers and reviews on the benefits of enriched environments. Here are a few I found helpful: Mark P. Mattson et al., "Suppression of Brain Aging and Neurodegenerative Disorders by Dietary Restriction and Environmental Enrichment: Molecular Mechanisms," *Mechanisms of Ageing and Development* 122 (2001): 757–78. Jess Nithianantharajah and Anthony J. Hannan, "Enriched Environments, Experience-Dependent Plasticity and Disorders of the Nervous System," *Nature Reviews Neuroscience* 7 (2006): 697–709. Jennifer C. Bennett et al., "Long-Term Continuous, but Not Daily, Environmental Enrichment Reduces Spatial Memory Decline in Aged Male Mice," *Neurobiology of Learning and Memory* 85 (2006): 139–52. N. Benaroya-Milshtein et al., "Environmental Enrichment in Mice Decreases Anxiety, Attenuates Stress Responses and Enhances Natural Killer Cell Activity," *European Journal of Neuroscience* 20 (2004): 1341–47. Lei Cao et al., "Environmental and Genetic Activation of a Brain-Adipocyte BDNF/Leptin Axis Causes Cancer Remission and Inhibition," *Cell* 142 (2010): 52–64. Agnieszka Z. Burzynska and Laura H. Malinin, "Enriched Environments for Healthy Aging: Qualities of

Seniors Housing Designs Promoting Brain and Cognitive Health," *Seniors Housing and Care Journal* 25 (2017): 15–37.

8 *one theory is that urban living*: Burzynska and Malinin, "Enriched Environments," 20.

8 *In her own research*: Laura Malinin, interview by author, November 19, 2018.

9 *Consider what happened*: Information about this experiment: Lynda M. D'Alessio, "The Impact of Neonatal ICU Single Family Rooms on Select Developmental Outcomes" (major paper, Rhode Island College School of Nursing, 2011), https://digitalcommons.ric.edu/etd/204/. Barry M. Lester et al., "Single-Family Room Care and Neurobehavioral and Medical Outcomes in Preterm Infants," *Pediatrics* 134 (2014): 754–60.

10 *noise can derail*: Gemma Brown, "NICU Noise and the Preterm Infant," *Neonatal Network* 28 (2009): 165–73. Elisha M. Wachman and Amir Lahav, "The Effects of Noise on Preterm Infants in the NICU," *Archives of Disease in Childhood: Fetal and Neonatal* 96 (2011): F305–309.

10 *Part of what made*: Lester et al., "Single-Family Room Care," 754–60. Barry M. Lester et al., "18-Month Follow-Up of Infants Cared For in a Single-Family Room Neonatal Intensive Care Unit," *Journal of Pediatrics* 177 (2016): 84–89.

1. THE INDOOR JUNGLE

12 *According to the latest estimates*: Ron Sender, Shai Fuchs, and Ron Milo, "Revised Estimates for the Number of Human and Bacteria Cells in the Body," *PLOS Biology* 14 (2016): e1002533.

13 *when a jet of hot water*: Noah Fierer, interview by author, September 21, 2016.

14 *Fierer found his calling*: Information on Fierer's background and career trajectory is from Noah Fierer, interview by author, January 18, 2017.

15 *In 2010, Fierer made*: Gilberto E. Flores et al., "Microbial Biogeography of Public Restroom Surfaces," *PLOS ONE* 6 (2011): e28132.

15 *The following year*: Gilberto E. Flores et al., "Diversity, Distribution and Sources of Bacteria in Residential Kitchens," *Environmental Microbiology* 15 (2013): 588–96.

15 *They began with a small pilot*: Information on the methods and results of this study comes from: Robert R. Dunn et al., "Home Life: Factors Structuring the Bacterial Diversity Found Within and Between Homes," *PLOS ONE* 8 (2013): e64133. Noah Fierer, interview by author, September 30, 2015. Rob Dunn, interview by author, September 29, 2015.

15 *So Fierer and Dunn launched*: Information on the methods and results of

this study comes from: Albert Barberán et al., "The Ecology of Microscopic Life in Household Dust," *Proceedings of the Royal Society B* 282 (2015): 20151139. Fierer, interview, September 30, 2015. Dunn, interview, September 29, 2015.

16 *In total, the indoor dust*: Noah Fierer, email to author, October 5, 2015.

16 *There are fewer than 25,000*: Rob Dunn, *Never Home Alone* (New York: Basic Books, 2018): 104.

16 *Our individual microbiomes*: Noah Fierer et al., "Forensic Identification Using Skin Bacterial Communities," *PNAS* 107 (2010): 6477–81. Eric A. Franzosa et al., "Identifying Personal Microbiomes Using Metagenomic Codes," *PNAS* 112 (2015): E2930–38.

16 *In one innovative study*: Simon Lax et al., "Longitudinal Analysis of Microbial Interaction between Humans and the Indoor Environment," *Science* 345 (2014): 1048–52. Jack Gilbert, interview by author, September 22, 2016.

17 *Fierer and Dunn found*: Information on sex differences is from: Barberán et al., "The Ecology of Microscopic Life." Fierer, interview, September 30, 2015. Dunn, interview, September 29, 2015.

17 *In a subsequent study*: J. C. Luongo et al., "Microbial Analyses of Airborne Dust Collected from Dormitory Rooms Predict the Sex of Occupants," *Indoor Air* 27 (2017): 338–44.

17 *Meanwhile, dogs introduce*: Information on how pets affect a home's microbes is from: Barberán et al., "The Ecology of Microscopic Life." Fierer, interview, September 30, 2015. Dunn, interview, September 29, 2015.

18 *Fungi are much less abundant*: Information on fungal patterns in homes is from: Barberán et al., "The Ecology of Microscopic Life." Fierer, interview, September 30, 2015. Dunn, interview, September 29, 2015.

18 *The geographic correlation was so strong*: Neal S. Grantham et al., "Fungi Identify the Geographic Origin of Dust Samples," *PLOS ONE* 10 (2015): e0122605.

18 *Fierer and Dunn did identify . . . fungi that live on human skin*: Barberán et al., "The Ecology of Microscopic Life." Dunn, *Never Home Alone*, 102–107. Dunn, interview, September 29, 2015.

18 *There are black yeasts*: P. Zalar et al., "Dishwashers: A Man-Made Ecological Niche Accommodating Human Opportunistic Fungal Pathogens," *Fungal Biology* 115 (2011): 997–1007. Maximilian Mora et al., "Resilient Microorganisms in Dust Samples of the International Space Station: Survival of the Adaptation Specialists," *Microbiome* 4 (2016): 65. Charles E. Robertson et al., "Culture-Independent Analysis of Aerosol Microbiology in a Metropolitan Subway System," *Applied and Environmental Microbiology* 79 (2013): 3485–93.

19 *When a group of BioBE researchers*: Information on the methods and results of this study comes from Steven W. Kembel et al., "Architectural Design Drives the Biogeography of Indoor Bacterial Communities," *PLOS ONE* 9 (2014): e87093.

19 *sunlight shining in through windows can inhibit*: Ashkaan K. Fahimipour et al., "Daylight Exposure Modulates Bacterial Communities Associated with Household Dust," *Microbiome* 6 (2018): 175.

19 *rooms that have operable*: Kembel et al., "Architectural Design Drives." Steve W. Kembel et al., "Architectural Design Influences the Diversity and Structure of the Built Environment Microbiome," *The ISME Journal* 6 (2012): 1469–79.

19 *In a 2016 study*: Jean F. Ruiz-Calderon et al., "Walls Talk: Microbial Biogeography of Homes Spanning Urbanization," *Science Advances* 2 (2016): e1501061.

20 *it can change as*: J. B. Emerson et al., "High Temporal Variability in Airborne Bacterial Diversity and Abundance Inside Single-Family Residences," *Indoor Air* 27 (2017): 576–86.

20 *Dampness can encourage*: Joanne E. Sordillo et al., "Home Characteristics as Predictors of Bacterial and Fungal Microbial Biomarkers in House Dust," *Environmental Health Perspectives* 119 (2011): 189–95.

20 *a rigorous cleaning regimen*: Flores et al., "Diversity, Distribution." Dunn et al., "Home Life." P. Rusin, P. Orosz-Coughlin, and C. Gerba, "Reduction of Faecal Coliform, Coliform and Heterotrophic Plate Count Bacteria in the Household Kitchen and Bathroom by Disinfection with Hypochlorite Cleaners," *Journal of Applied Microbiology* 85 (1998): 819–28. A. Medrano-Félix et al., "Impact of Prescribed Cleaning and Disinfectant Use on Microbial Contamination in the Home," *Journal of Applied Microbiology* 110 (2011): 463–71.

20 *as Dunn himself learned*: Information on the methods and results of this study comes from: Mary Jane Epps et al., "Too Big to Be Noticed: Cryptic Invasion of Asian Camel Crickets in North American Houses," *PeerJ* 2 (2014): e523. Dunn, *Never Home Alone*, 120–27. Rob Dunn, interview by author, February 14, 2018.

21 *In 2012, Dunn's team*: Information on the methods and results of this study comes from: Matthew A. Bertone et al., "Arthropods of the Great Indoors: Characterizing Diversity Inside Urban and Suburban Homes," *PeerJ* 19 (2016): e1582. Dunn, *Never Home Alone*, 128–37. Dunn, interview, February 14, 2018.

21 *He and Fierer made*: Anne A. Madden et al., "The Diversity of Arthropods in Homes Across the United States as Determined by Environmental DNA

Analyses," *Molecular Ecology* 25 (2016): 6214–24. Fierer, interview, September 21, 2016. Dunn, interview, February 14, 2018.

22 *Bedbugs and German cockroaches*: Martin et al., "Evolution of the Indoor Biome."

22 *the black yeasts*: Niichiro Abe and Nobuo Hamada, "Molecular Characterization and Surfactant Utilization of *Scolecobasidium* Isolates from Detergent-Rich Indoor Environments," *Biocontrol Science* 16 (2011): 139–47.

22 *Fierer recently discovered*: Karyna Rosario et al., "Diversity of DNA and RNA Viruses in Indoor Air as Assessed via Metagenomic Sequencing," *Environmental Science and Technology* 52 (2018): 1014–27.

22 *Indoor microbes, insects, and rodents*: Martin et al., "Evolution of the Indoor Biome."

22 *German cockroaches are known*: Ayako Wada-Katsumata, Jules Silverman, and Coby Schal, "Changes in Taste Neurons Support the Emergence of an Adaptive Behavior in Cockroaches," *Science* 340 (2013): 972–75.

22 *Some indoor insects*: Martin et al., "Evolution of the Indoor Biome."

22 *Dunn and other ecologists*: Martin et al., "Evolution of the Indoor Biome."

22 *Fierer and Dunn found*: Barberán et al., "The Ecology of Microscopic Life."

22 *These pests can have*: Snehal N. Shah et al., "Housing Quality and Mental Health: The Association Between Pest Infestation and Depressive Symptoms Among Public Housing Residents," *Journal of Urban Health* 95 (2018): 691–702.

23 *In Thailand, house spiders*: Dunn, *Never Home Alone*, 294.

23 *Some African and Latin American*: Dunn, *Never Home Alone*, 172.

23 *Homes in the Pacific Northwest*: Joseph M. Craine et al., "Molecular Analysis of Environmental Plant DNA in House Dust Across the United States," *Aerobiologia* 33 (2017): 71–86.

23 *In one recent experiment*: Marina Vance, "Homes Harbor Airborne Chemicals" (press briefing, AAAS Annual Meeting, Washington, DC, February 17, 2019).

23 *Although houseplants*: Michael S. Waring, "Bio-walls and Indoor Houseplants: Facts and Fictions" (presentation, Microbiomes of the Built Environment: From Research to Application, Meeting #3, Irvine, CA, October 17–18, 2016), http://nas-sites.org/builtmicrobiome/files/2016/07/Michael-Waring-FOR-POSTING.pdf.

24 Aspergillus fumigatus, *a fungus*: "Pillows: A Hot Bed of Fungal Spores," *EurekAlert!*, October 14, 2005, www.eurekalert.org/pub_releases/2005-10/uom-p-a101305.php.

24 Streptococcus *bacteria*: Barberán et al., "The Ecology of Microscopic Life."

24 *Some scientists have theorized*: The hygiene hypothesis was first proposed in D. P. Strachan, "Hay Fever, Hygiene, and Household Size," *BMJ* 299 (1989): 1259–60. It is also sometimes known as the "old friends" hypothesis, as outlined in Graham A. Rook, "Regulation of the Immune System by Biodiversity from the Natural Environment: An Ecosystem Service Essential to Health," *PNAS* 110 (2013): 18360–67.

25 *Studies show that children*: Kei E. Fujimura et al., "Man's Best Friend? The Effect of Pet Ownership on House Dust Microbial Communities," *Journal of Allergy and Clinical Immunology* 126 (2010): 410–12. Dennis R. Ownby et al., "Exposure to Dogs and Cats in the First Year of Life and Risk of Allergic Sensitization at 6 to 7 Years of Age," *JAMA* 288 (2002): 963–72. Tove Fall et al., "Early Exposure to Dogs and Farm Animals and the Risk of Childhood Asthma," *JAMA Pediatrics* 169 (2015): e153219.

25 *Children who grow up on farms*: Fall et al., "Early Exposure to Dogs." Erika von Mutius and Donata Vercelli, "Farm Living: Effects on Childhood Asthma and Allergies," *Nature Reviews Immunology* 10 (2010): 861–68.

25 *Although the groups have much in common*: Information on the similarities and differences between the Amish and the Hutterites comes from Michelle M. Stein et al., "Innate Immunity and Asthma Risk in Amish and Hutterite Farm Children," *New England Journal of Medicine* 375 (2016): 411–21.

25 *In 2016, scientists reported*: Information on the methods and results of this study comes from Stein et al., "Innate Immunity."

26 *which has been found in dog noses*: Barbara Tress et al., "Bacterial Microbiome of the Nose of Healthy Dogs and Dogs with Nasal Disease," *PLOS ONE* 12 (2017): e0176736. Claudia Schabereiter-Gurtner et al., "Phylogenetic Diversity of Bacteria Associated with Paleolithic Paintings and Surrounding Rock Walls in Two Spanish Caves (Llonín and La Garma)," *FEMS Microbiology Ecology* 47 (2004): 235–47.

27 *which are becoming more common*: Matthew J. Gebert et al., "Ecological Analyses of Mycobacteria in Showerhead Biofilms and Their Relevance to Human Health," *mBio* 9 (2018): e01614–18. Fierer, interview, September 21, 2016.

27 *In fact, Fierer and Dunn*: Gebert et al., "Ecological Analyses of Mycobacteria." Noah Fierer, phone interview by author, November 13, 2017.

27 *Fierer and Dunn did discover*: Gebert et al., "Ecological Analyses of Mycobacteria."

28 *"In the future"*: National Academies of Sciences, Engineering, and Medicine, *Microbiomes of the Built Environment: A Research Agenda for Indoor Micro-*

biology, Human Health, and Buildings (Washington, DC: National Academies Press, 2017), 109.

28 *In one alarming study*: B. Andersen et al., "Pre-contamination of New Gypsum Wallboard with Potentially Harmful Fungal Species," *Indoor Air* 27 (2017): 6–12. Also discussed in Dunn, *Never Home Alone*, 110–13.

28 *boosting the rate of air ventilation*: National Academies, *Microbiomes of the Built Environment*, 152–55.

28 *Removing carpeting*: National Academies, *Microbiomes of the Built Environment*, 101. Jing Qian, Jordan Peccia, and Andrea R. Ferro, "Walking-Induced Particle Resuspension in Indoor Environments," *Atmospheric Environment* 89 (2014): 464–81. Denina Hospodsky et al., "Human Occupancy as a Source of Indoor Airborne Bacteria," *PLOS ONE* 7 (2012): e34867.

29 *Hartmann has discovered*: Erica M. Hartmann et al., "Antimicrobial Chemicals Are Associated with Elevated Antibiotic Resistance Genes in the Indoor Dust Microbiome," *Environmental Science and Technology* 50 (2016): 9807–15. Ashkaan K. Fahimipour et al., "Antimicrobial Chemicals Associate with Microbial Function and Antibiotic Resistance Indoors," *mSystems* 3 (2018): e00200–18.

29 *The manufacturer of one*: "Homebiotic," accessed April 30, 2019, www .homebiotic.com/.

29 *But few of these products*: National Academies, *Microbiomes of the Built Environment*, 161.

2. A HOSPITAL ROOM OF ONE'S OWN

31 *a yearlong study of microbes in the new hospital*: Simon Lax et al., "Bacterial Colonization and Succession in a Newly Opened Hospital," *Science Translational Medicine* 9 (2017): eaah6500.

32 *SARS, the deadly respiratory virus*: Carolyn Farquharson and Karen Baguley, "Responding to the Severe Acute Respiratory Syndrome (SARS) Outbreak: Lessons Learned in a Toronto Emergency Department," *Journal of Emergency Nursing* 29 (2003): 222–28.

32 *when one hospitalized patient*: M. K. Shaughnessy et al., "Evaluation of Hospital Room Assignment and Acquisition of *Clostridium difficile* Infection," *Infection Control and Hospital Epidemiology* 32 (2011): 201–206.

32 *which affect 7 to 10 percent*: World Health Organization, "Health Care–Associated Infections: FACT SHEET," accessed August 2, 2019, www.who .int/gpsc/country_work/burden_hcai/en/.

32 *When administrators at*: Information on the redesign and its results comes from: Torsten Holmdahl and Peter Lanbeck, "Design for the Post-Antibiotic

Era: Experiences from a New Building for Infectious Diseases in Malmö, Sweden," *HERD* 6 (2013): 27–52. Torsten Holmdahl, "Hospital Design and Room Decontamination for a Post-Antibiotic Era and an Era of Emerging Infectious Diseases" (PhD diss., Lund University, 2017). Torsten Holmdahl, interview by author, January 22, 2019.

33 *When Montreal General Hospital*: Dana Y. Teltsch et al., "Infection Acquisition Following Intensive Care Unit Room Privatization," *JAMA Internal Medicine* 171 (2011): 32–38.

33 *providing conveniently located sinks*: Roger S. Ulrich et al., "A Review of the Research Literature on Evidence-Based Healthcare Design," *HERD* 1 (2008): 61–125.

34 *the right design decision*: Ulrich et al., "A Review of the Research."

34 *The modern hospital is*: Information on the history of hospitals is from: Guenter B. Risse, *Mending Bodies, Saving Souls: A History of Hospitals* (New York: Oxford University Press, 1999). Sethina Watson, "The Origins of the English Hospital," *Transactions of the Royal Historical Society* 16 (2006): 75–94. Nick Black, "Rise and Demise of the Hospital: A Reappraisal of Nursing," *BMJ* 331 (2005): 1394–96. Jeanne Susan Kisacky, "Restructuring Isolation: Hospital Architecture, Medicine, and Disease Prevention," *Bulletin of the History of Medicine* 79 (2005): 1–49. Tom Gormley, "The History of Hospitals and Wards," *Healthcare Design* 10 (2010): 50–54. Barbra Mann Wall, "History of Hospitals," University of Pennsylvania School of Nursing, accessed April 28, 2019, www.nursing.upenn.edu/nhhc/nurses-institutions -caring/history-of-hospitals/.

34 *These were the appalling conditions*: Information on Florence Nightingale's work during the Crimean War is from: Christopher J. Gill and Gillian C. Gill, "Nightingale in Scutari: Her Legacy Reexamined," *Clinical Infectious Diseases* 40 (2005): 1799–1805. Warren Winkelstein Jr., "Florence Nightingale: Founder of Modern Nursing and Hospital Epidemiology," *Epidemiology* 20 (2009): 311. Maya Aravind and Kevin C. Chung, "Evidence-Based Medicine and Hospital Reform: Tracing Origins Back to Florence Nightingale," *Plastic and Reconstructive Surgery* 125 (2010): 403–409. Craig Zimring and Jennifer DuBose, "Healthy Health Care Settings," in *Making Healthy Places*, ed. Andrew L. Dannenberg, Howard Frumkin, and Richard J. Jackson (Washington, DC: Island Press, 2011): 205.

35 *She advocated for*: Florence Nightingale, *Notes on Hospitals* (London: John W. Parker, 1859).

35 *"To shut up your patients"*: Nightingale, *Notes on Hospitals*, 16.

35 *"Among kindred effects"*: Nightingale, *Notes on Hospitals*, 19.

35 *Nightingale endorsed*: Nightingale, *Notes on Hospitals*. Aravind and Chung, "Evidence-Based Medicine."

36 *in which long, skinny*: Information on pavilion-style hospitals is from: Nightingale, *Notes on Hospitals*. Anthony King, "Hospital Planning: Revised Thoughts on the Origin of the Pavilion Principle in England," *Medical History* 10 (1966): 360–73. G. C. Cook, "Henry Currey FRIBA (1820–1900): Leading Victorian Hospital Architect, and Early Exponent of the 'Pavilion Principle,'" *Postgraduate Medical Journal* 78 (2002): 352–59. Cynthia Imogen Hammond, "Reforming Architecture, Defending Empire: Florence Nightingale and the Pavilion Hospital," in *Un/Healthy Interiors: Contestations at the Intersection of Public Health and Private Space*, ed. Aran S. MacKinnon and Jonathan D. Ablard (Studies in the Social Sciences, University of West Georgia, 2005): 1–24. Zimring and DuBose, "Healthy Health Care Settings," 205.

36 *As germ theory*: Information on how hospital design evolved during the twentieth century: C. Robert Horsburgh, "Healing by Design," *The New England Journal of Medicine* 333 (1995): 735–40. Angela Burke, "Towards a New Hospital Architecture: An Exploration of the Relationship Between Hospital Space and Technology" (PhD diss., University of East London, 2014). Zimring and DuBose, "Healthy Health Care Settings," 205–206.

36 *interviewing dozens of Ann Arbor*: Information on the methods and results of this study comes from: Roger S. Ulrich, "Scenery and the Shopping Trip: The Roadside Environment as a Factor in Route Choice" (PhD diss., University of Michigan, 1974).

36n *Hospitals in low-income*: World Health Organization and the United Nations Children's Fund, *WASH in Health Care Facilities: Global Baseline Report 2019* (Geneva: WHO and UNICEF, 2019).

37 *For a study he published*: Information on the methods and results of this study comes from: Roger S. Ulrich, "Visual Landscapes and Psychological Well-Being," *Landscape Research* 4 (1979): 17–23.

37 *He had been a sickly kid*: Information on Ulrich's childhood experiences is from Roger Ulrich, interview by author, December 14, 2016.

38 *the perfect setting for his study*: Information on the methods and results of this study comes from: Roger S. Ulrich, "View Through a Window May Influence Recovery from Surgery," *Science* 224 (1984): 420–21.

39 *The first was*: Ann Sloan Devlin, interview by author, May 10, 2018. Dak Kopec, *Environmental Psychology for Design* (New York: Fairchild Publications, 2006): 211–12. Horsburgh, "Healing by Design."

39 *The second was*: Devlin, interview, May 10, 2018. Zimring and DuBose,

"Healthy Health Care Settings," 206. Aravind and Chung, "Evidence-Based Medicine."

39 *In the early 1990s*: Roger S. Ulrich, "Effects of Interior Design on Wellness: Theory and Recent Scientific Research," *Journal of Health Care Interior Design* 3 (1991): 104. Ulrich, "A Review of the Research Literature," 129.

39 *Other researchers found*: Gregory B. Diette et al., "Distraction Therapy with Nature Sights and Sounds Reduces Pain During Flexible Bronchoscopy," *Chest* 123 (2003): 941–48. A. C. Miller, L. C. Hickman, and G. K. Lemasters, "A Distraction Technique for Control of Burn Pain," *The Journal of Burn Care and Rehabilitation* 13 (1992): 576–80.

39 *Indoor plants*: Seong-Hyun Park and Richard H. Mattson, "Effects of Flowering and Foliage Plants in Hospital Rooms on Patients Recovering from Abdominal Surgery," *HortTechnology* 18 (2008): 563–68. Seong-Hyun Park and Richard H. Mattson, "Ornamental Indoor Plants in Hospital Rooms Enhanced Health Outcomes of Patients Recovering from Surgery," *The Journal of Alternative and Complementary Medicine* 15 (2009): 975–80.

39 *So natural settings*: Ulrich, "A Review of the Research Literature." Ulrich, "Effects of Interior Design." Ulrich, interview, December 14, 2016.

40 *In a series of studies*: Q. Li et al., "Visiting a Forest, but Not a City, Increases Human Natural Killer Activity and Expression of Anti-cancer Proteins," *International Journal of Immunopathology and Pharmacology* 21 (2008): 117–27.

40 *They use fewer painkillers*: Jeffrey Walch et al., "The Effect of Sunlight on Postoperative Analgesic Medication Use: A Prospective Study of Patients Undergoing Spinal Surgery," *Psychosomatic Medicine* 67 (2005): 156–63. Kathleen M. Beauchemin and Peter Hays, "Sunny Hospital Rooms Expedite Recovery from Severe and Refractory Depressions," *Journal of Affective Disorders* 40 (1996): 49–51. Kathleen M. Beauchemin and Peter Hays, "Dying in the Dark: Sunshine, Gender and Outcomes in Myocardial Infarction," *Journal of the Royal Society of Medicine* 91 (1998): 352–54.

40 *Though it's difficult*: The benefits of daylight for hospital patients are reviewed in Ulrich, "A Review of the Research Literature."

40 *wards that keep the lights on*: Ulrich, "A Review of the Research Literature."

41 *Hospitals can be as noisy*: Roger Ulrich et al., *The Role of the Physical Environment in the Hospital of the 21st Century: A Once-in-a-Lifetime Opportunity* (Concord, CA: Center for Health Design, 2004).

41 *In 2002, Ulrich*: Information on the methods and results of this study comes from: Ulrich, interview, December 14, 2016. Inger Hagerman et al., "Influence of Intensive Coronary Care Acoustics on the Quality of Care and Physiological State of Patients," *International Journal of Cardiology* 98 (2005): 267–70. V. Blomkvist et al., "Acoustics and Psychosocial Environment in

Intensive Coronary Care," *Occupational and Environmental Medicine* 62 (2005): e1. Roger Ulrich, e-mail to author, April 30, 2019.

41 *In addition to reducing infections*: There are some good reviews on the benefits of single-patient rooms in: Ulrich et al., *The Role of the Physical Environment*. Habib Chaudhury, Atiya Mahmood, and Maria Valente, "Advantages and Disadvantages of Single- Versus Multiple-Occupancy Rooms in Acute Care Environments: A Review and Analysis of the Literature," *Environment and Behavior* 37 (2005): 760–86.

41 *One study suggests that emergency room*: David Barlas et al., "Comparison of the Auditory and Visual Privacy of Emergency Department Treatment Areas with Curtains Versus Those with Solid Walls," *Annals of Emergency Medicine* 38 (2001): 135–39.

42 *Called "acuity-adaptable" rooms*: Ann L. Hendrich, Joy Fay, and Amy K. Sorrells, "Effects of Acuity-Adaptable Rooms on Flow of Patients and Delivery of Care," *American Journal of Critical Care* 13 (2004): 35–45.

42 *In 2004, a group*: Information on the Fable Hospital is from: Leonard L. Berry et al., "The Business Case for Better Buildings," *Frontiers of Health Services Management* 21 (2004): 3–24. Derek Parker, interview by author, February 23, 2017.

42 *In 2004, the same year*: Information on Dublin Methodist is from: Cheryl Herbert, "Case Study: Dublin Methodist Hospital," *The Hastings Center Report* 41 (2011): 23–24. Cheryl Herbert, interview by author, March 8, 2017.

43 *In the first few years*: Herbert, "Case Study," 24.

44 *Design guidelines issued*: *Guidelines for Design and Construction of Health Care Facilities* (American Institute of Architects/Facilities Guidelines Institute, 2006).

44 *As surgical practices*: Information on operating room activities and design is from: Burke, "Towards a New Hospital Architecture." Anjali Joseph et al., "Safety, Performance, and Satisfaction Outcomes in the Operating Room: A Literature Review," *HERD* 11 (2018): 137–50. Anjali Joseph, interview by author, September 26, 2016. Anjali Joseph, interview by author, September 25, 2018. David Allison, interview by author, September 26, 2018.

45 *in developed nations*: *WHO Guidelines for Safe Surgery 2009* (Geneva: World Health Organization, 2009), 2.

45 *The four-year project*: *Realizing Improved Patient Care through Human-Centered Design in the Operating Room*, vol. 1 (Clemson Center for Health Facilities Design and Testing, 2016), 1–3, https://issuu.com/clemsonchfdt /docs/ripchd.or_volume_1. Joseph, interview, September 26, 2016.

46 *To do so, they recorded*: Anjali Joseph et al., "Minor Flow Disruptions, Traffic-Related Factors and Their Effect on Major Flow Disruptions in

the Operating Room," *BMJ Quality and Safety* 28 (2019): 276–83. Sara Bayramzadeh et al., "The Impact of Operating Room Layout on Circulating Nurse's Work Patterns and Flow Disruptions: A Behavioral Mapping Study," *HERD* 11 (2018): 124–38.

46 *These "flow disruptions"*: Joseph et al., "Minor Flow Disruptions." Douglas A. Wiegmann et al., "Disruptions in Surgical Flow and Their Relationship to Surgical Errors: An Exploratory Investigation," *Surgery* 142 (2007): 658–65.

46 *The researchers found*: The team's findings on flow disruptions are detailed in: Joseph et al., "Minor Flow Disruptions." Bayramzadeh et al., "The Impact of Operating Room Layout." David M. Neyens et al., "Using a Systems Approach to Evaluate a Circulating Nurse's Work Patterns and Workflow Disruptions," *Applied Ergonomics*, published ahead of print, March 30, 2018, https://linkinghub.elsevier.com/retrieve/pii/S0003-6870(18)30078-4. *Realizing Improved Patient Care Through Human-Centered Design in the Operating Room*, vol. 2 (Clemson Center for Health Facilities Design and Testing, 2017), 81, https://issuu.com/clemsonchfdt/docs/ripchd.or_volume_2. Sara Bayramzadeh, "Study Findings" (presentation, RIPCHD.OR Learning Lab Workshop, Charleston, SC, January 25, 2018). Kevin Taaffe, "Study Findings" (presentation, RIPCHD.OR Learning Lab Workshop, Charleston, SC, January 25, 2018).

47 *using rolls of colored tape*: David Allison, "Study Findings" (presentation, RIPCHD.OR Learning Lab Workshop, Charleston, SC, January 25, 2018). *Realizing Improved Patient Care*, vol. 2, 34–35.

47 *testing their hunches*: Information on the computer model and findings is from: Amin Khoshkenar et al., "Simulation-Based Design and Traffic Flow Improvements in the Operating Room," *Proceedings of the 2017 Winter Simulation Conference* (2017): 2975–83. *Realizing Improved Patient Care*, vol. 2, 82–83. Kevin Taaffe, interview by author, August 14, 2018. Allison, interview, September 26, 2018.

47 *Initially, they'd placed*: Information about how and why they repositioned the OR table is from: Allison, interview, September 26, 2018. Taaffe, interview, August 14, 2018. *Realizing Improved Patient Care*, vol. 2, 36–49.

47n *Moreover, limiting movement*: Kevin Taaffe et al., "The Influence of Traffic, Area Location, and Other Factors on Operating Room Microbial Load," *Infection Control and Hospital Epidemiology* 39 (2018): 391–97.

48 *They placed the circulating*: Other features of the OR prototype are from: *Realizing Improved Patient Care*, vol. 2, 36–49. Allison, "Study Findings." Allison, interview, September 26, 2018. Joseph, interview, September 26, 2016. Anjali Joseph, interview by author, September 25, 2018.

52 *In the following months*: David Allison, e-mail to author, March 24, 2019.

52 *When the first*: Joseph, interview, September 25, 2018.

52 *Joseph's team is monitoring*: Joseph, interview, September 25, 2018.

52 *They're considering applying*: Joseph, interview, September 25, 2018.

52 *In Philadelphia, for instance*: Bon Ku, interview by author, November 2, 2017.

3. STAIR MASTERS

55 *Yellow fever plagued*: Information on infectious disease in nineteenth-century New York is from: New York City Department of Health and Mental Hygiene, *Protecting Public Health in New York City: 200 Years of Leadership* (New York: Bureau of Communications, DOHMH, 2005), 4–10. Citizens' Association of New York, *Report of the Council of Hygiene and Public Health of the Citizens' Association of New York upon the Sanitary Condition of the City* (New York: D. Appleton and Company, 1865). Richard Plunz, *A History of Housing in New York City* (New York: Columbia University Press, 2016), 2–3.

55 *By 1863*: Citizens' Association, *Report of the Council of Hygiene*, x–xi.

55 *In some years*: New York City Department of Health and Mental Hygiene, *Protecting Public Health*, 10.

55 *Local civic groups*: Information on the city's squalor is from: Citizens' Association, *Report of the Council of Hygiene*. Department of Health and Mental Hygiene, *Protecting Public Health*, 7–10. Plunz, *A History of Housing*, 2–3, 13–16, 50–55.

56 *In 1865, the Citizens' Association*: Citizens' Association, *Report of the Council of Hygiene*, xi.

56 *In 1866, it created*: New York City Department of Health and Mental Hygiene, *Protecting Public Health*, 12–15.

56 *Fifteen years later*: "History," New York City Department of Sanitation, accessed February 8, 2018, www1.nyc.gov/assets/dsny/about/inside-dsny/history.shtml.

56 *Legislators passed*: Plunz, *A History of Housing*, 22–49. City of New York, *Active Design Guidelines: Promoting Physical Activity and Health in Design* (2010), 13.

57n *Housing quality and urban planning . . . dropped by more than 20 percent*: Carl-Johan Neiderud, "How Urbanization Affects the Epidemiology of Emerging Infectious Diseases," *Infection Ecology and Epidemiology* 5 (2015): 10.3402/iee.v5.27060. Mauricio L. Barreto et al., "Effect of City-wide Sanitation Programme on Reduction in Rate of Childhood Diarrhoea in Northeast Brazil: Assessment by Two Cohort Studies," *Lancet* 370 (2007): 1622–28.

58 *His administration began*: New York City Department of Health and Mental Hygiene, *Preventing Non-communicable Diseases and Injuries: Innovative Solutions from New York City* (New York: New York City Department of Health and Mental Hygiene, 2011), 6–7. Thomas R. Frieden et al., "Adult Tobacco Use Levels After Intensive Tobacco Control Measures: New York City, 2002–2003," *American Journal of Public Health* 95 (2005): 1016–23.

58 *Then officials turned*: New York City Department of Health and Mental Hygiene, *Preventing Non-communicable Diseases*, 6.

58 *New Yorkers were largely sedentary*: New York City Department of Health and Mental Hygiene, *Preventing Non-communicable Diseases*, 6. Emily N. Ussery et al., "Joint Prevalence of Sitting Time and Leisure-Time Physical Activity among US Adults, 2015–2016," *JAMA* 320 (2018): 2036–38.

58 *they ate too much*: New York City Department of Health and Mental Hygiene, *Preventing Non-communicable Diseases*, 6. U.S. Department of Health and Human Services and U.S. Department of Agriculture, *2015–2020 Dietary Guidelines for Americans* (2015), 38–39.

58 *And in New York*: NCD Risk Factor Collaboration, "Trends in Adult Body-Mass Index in 200 Countries from 1975 to 2014: A Pooled Analysis of 1698 Population-Based Measurement Studies with 19.2 Million Participants," *Lancet* 387 (2016): 1377–96.

58 *By 2004, nearly*: Gretchen Van Wye, "Obesity and Diabetes in NYC, 2002 and 2004," *Preventing Chronic Disease* 5 (2008): A48.

58 *In an effort to help*: New York City Department of Health and Mental Hygiene, *Preventing Non-communicable Diseases*, 9–12.

58n *Some of these moves*: Melecia Wright et al., "Impact of a Municipal Policy Restricting Trans Fatty Acid Use in New York City Restaurants on Serum Trans Fatty Acid Levels in Adults," *American Journal of Public Health* 4 (2019): 634–36. Eric J. Brandt et al., "Hospital Admissions for Myocardial Infarction and Stroke Before and After the Trans-Fatty Acid Restrictions in New York," *JAMA Cardiology* 2 (2017): 627–34. Maya K. Vadiveloo, L. Beth Dixon, and Brian Elbel, "Consumer Purchasing Patterns in Response to Calorie Labeling Legislation in New York City," *International Journal of Behavioral Nutrition and Physical Activity* 8 (2011): 51.

59 *where obesity, diabetes, and high blood pressure*: Jonathan B. Wallach and Mariano J. Rey, "A Socioeconomic Analysis of Obesity and Diabetes," *Preventing Chronic Disease* 6 (2009): A108. Jennifer L. Black et al., "Neighborhoods and Obesity in New York City," *Health and Place* (2010): 489–99. New York City Department of Health and Mental Hygiene, "Diabetes in New York City," Epi Data Brief 26 (2013), www1.nyc.gov/site/doh/data/data-sets/epi-data-briefs-and-data-tables.page. New York City Department of

Health and Mental Hygiene, "Hypertension in New York City: Disparities in Prevalence," Epi Data Brief 82 (2016), www1.nyc.gov/site/doh/data/data-sets/epi-data-briefs-and-data-tables.page.

59 *So the city doled out*: Paul M. Kelly, "Obesity Prevention in a City State: Lessons from New York City during the Bloomberg Administration," *Frontiers in Public Health* 4 (2016): 60. Margaret Leggat et al., "Pushing Produce: The New York City Green Carts Initiative," *Journal of Urban Health* 89 (2012): 937–38.

60 *Adults who live*: Reid Ewing, "Relationship Between Urban Sprawl and Physical Activity, Obesity, and Morbidity," *American Journal of Health Promotion* 18 (2003): 47–57.

60 *Scientists have found*: Brian E. Saelens, James F. Sallis, and Lawrence D. Frank, "Environmental Correlates of Walking and Cycling: Findings from the Transportation, Urban Design, and Planning Literatures," *Annals of Behavioral Medicine* 25 (2003): 80–91. Brian E. Saelens and Susan L. Handy, "Built Environment Correlates of Walking: A Review," *Medicine and Science in Sports and Exercise* 40 (2008): S550–66.

60 *and that those who live*: Bahman P. Tabaei et al., "Associations of Residential Socioeconomic, Food, and Built Environments with Glycemic Control in Persons with Diabetes in New York City from 2007–2013," *American Journal of Epidemiology* 187 (2018): 736–45. Chinmoy Sarkar, Chris Webster, and John Gallacher, "Neighbourhood Walkability and Incidence of Hypertension: Findings from the Study of 429,334 UK Biobank Participants," *International Journal of Hygiene and Environmental Health* 221 (2018): 458–68.

60 *New York City residents*: Andrew Rundle et al., "The Urban Built Environment and Obesity in New York City: A Multilevel Analysis," *American Journal of Health Promotion* 21 (2007): 326–34.

60 *And low-income neighborhoods*: Latetia V. Moore et al., "Availability of Recreational Resources in Minority and Low Socioeconomic Status Areas," *American Journal of Preventive Medicine* 34 (2008): 16–22. Paul A. Estabrooks, Rebecca E. Lee, and Nancy C. Gyurcsik, "Resources for Physical Activity Participation: Does Availability and Accessibility Differ by Neighborhood Socioeconomic Status?" *Annals of Behavioral Medicine* 25 (2003): 100–104. Simon D. S. Fraser and Karen Lock, "Cycling for Transport and Public Health: A Systematic Review of the Effect of the Environment on Cycling," *European Journal of Public Health* 21 (2011): 738–43. Deborah A. Cohen, "Contribution of Public Parks to Physical Activity," *American Journal of Public Health* 97 (2007): 509–14.

60 *In large, longitudinal*: H. D. Sesso et al., "Physical Activity and Cardiovascular Disease Risk in Middle-Aged and Older Women," *American Journal*

of Epidemiology 150 (1999): 408–16. Ralph S. Paffenbarger et al., "The Association of Changes in Physical-Activity Level and Other Lifestyle Characteristics with Mortality Among Men," *New England Journal of Medicine* 328 (1993): 538–45.

61 *In 2006*: David Burney, interview by author, February 19, 2018. American Institute of Architects New York Chapter, *Fit-City: Promoting Physical Activity Through Design* (New York: AIA New York, 2006), www.aiany.org /wp-content/uploads/2017/10/FitCity1_Publication_Final_162.pdf.

61 *The guidelines, published in 2010*: City of New York, *Active Design Guidelines.*

61 *The guidelines also urge*: City of New York, *Active Design Guidelines*, 70–77. Gayle Nicoll, "Spatial Measures Associated with Stair Use," *American Journal of Health Promotion* 21 (2007): 346–52. Ryan R. Ruff et al., "Associations Between Building Design, Point-of-Decision Stair Prompts, and Stair Use in Urban Worksites," *Preventive Medicine* 60 (2014): 60–64. David R. Bassett et al., "Architectural Design and Physical Activity: An Observational Study of Staircase and Elevator Use in Different Buildings," *Journal of Physical Activity and Health* 10 (2013): 556–62.

61 *Installing signs encouraging*: City of New York, *Active Design Guidelines*, 78–79. Ruff et al., "Associations Between Building Design." Marc Nocon et al., "Increasing Physical Activity with Point-of-Choice Prompts: A Systematic Review," *Scandinavian Journal of Public Health* 38 (2010): 633–38. Kerri N. Boutelle et al., "Using Signs, Artwork, and Music to Promote Stair Use in a Public Building," *American Journal of Public Health* 91 (2001): 2004–2006. Karen K. Lee et al., "Promoting Routine Stair Use," *American Journal of Preventive Medicine* 42 (2012): 136–41. Nicole Angelique Kerr et al., "Increasing Stair Use in a Worksite through Environmental Changes," *American Journal of Health Promotion* 18 (2004): 312–15.

62 *In 2007, the Geneva University Hospitals*: Philippe Meyer et al., "Stairs Instead of Elevators at the Workplace: Cardioprotective Effects of a Pragmatic Intervention," *European Journal of Preventive Cardiology* 17 (2010): 569–75.

62 *In 2013, Bloomberg issued*: Office of the Mayor, "Mayor Bloomberg Announces First Ever Center for Active Design to Promote Physical Activity and Health in Buildings and Public Spaces Through Building Code and Design Standard Changes," news release, July 17, 2013, www1.nyc.gov/office-of-the-mayor /news/250-13/mayor-bloomberg-first-ever-center-active-design-promote -physical-activity-and.

63 *in a Bronx neighborhood where*: Information about the neighborhood and Arbor House is from: Elizabeth Garland et al., "Active Design in Affordable Housing: A Public Health Nudge," *Preventive Medicine Reports* 10

(2018): 9–14. Elizabeth Garland et al., "One Step at a Time Towards Better Health: Active Design in Affordable Housing," *Environmental Justice* 7 (2014): 166–71. "Blue Sea Development Company," Center for Active Design, accessed May 8, 2019, https://centerforactivedesign.org/awards/blueseadevelopmentcompany.

63 *in a focus group*: Garland et al., "One Step at a Time."

64 *The Mount Sinai team tracked*: Garland et al., "Active Design in Affordable Housing."

64 *the Robert Wood Johnson Foundation*: Transtria LLC, *Evaluation of Active Living by Design: Learning from 25 Community Partnerships* (Robert Wood Johnson Foundation, 2012), www.transtria.com/pdfs/ALbD/Cross_site _assessment.pdf.

64 *The private sector*: "BlueCross BlueShield of Tennessee," Center for Active Design, accessed May 8, 2019, https://centerforactivedesign.org /bluecrossblueshield. "Google," Center for Active Design, accessed May 8, 2019, https://awards.centerforactivedesign.org/winners/google.

64 *Kids spend as many as half*: Jeri Brittin et al., "Physical Activity Design Guidelines for School Architecture," *PLOS ONE* 10 (2015): e0132597.

65 *He laid out*: Nicholas Gorman et al., "Designer Schools: The Role of School Space and Architecture in Obesity Prevention," *Obesity* 15 (2007): 2521–30.

65 *By 2009, Dillwyn Primary School*: Pennie Allen, interview by author, September 5, 2017. Pennie Allen, interview by author, April 17, 2018.

66 *in the nineteenth century*: "Buckingham County, Virginia," Division of Geology and Mineral Resources, accessed May 8, 2019, www.dmme.virginia .gov/dgmr/buckingham.shtml.

66 *More than 20 percent*: "Buckingham County, VA," Census Reporter, accessed May 8, 2019, https://censusreporter.org/profiles/05000US51029-buckingham -county-va/.

66 *70 percent*: Leah Frerichs et al., "Children's Discourse of Liked, Healthy, and Unhealthy Foods," *Journal of the Academy of Nutrition and Dietetics* 116 (2016): 1323–31.

66 *They tend to struggle*: Allen, interview, September 5, 2017.

66 *Despite the county's bucolic setting*: Allen, interview, April 17, 2018. Terry Huang and Matthew Trowbridge, "Insights 3—Healthy by Design: Architecture's New Terrain; The Dining Commons: Promoting Healthy and Active FoodSmart™ Kids," 4, http://wemoveschoolsforward.com/wp-content /uploads/2017/05/Buckingham_Insights_Case_Study.pdf.

67 *The team decided that*: Information about the new schools and their features and programs is from a variety of sources, including my own visit to Buckingham on March 29, 2017, and the numerous publications available at VMDO's

"We Move Schools Forward" website (http://wemoveschoolsforward.com /publications/). Other sources for this material include: Dina Sorensen, interview by author, October 25, 2016. Dina Sorensen, interview by author, March 29, 2017. Dina Sorensen, "Evidence-Based Design: Lessons from Virginia" (presentation, FitKids Symposium: Designing Spaces for Play, New York, NY, October 19, 2017). Dina Sorensen, interview by author, January 31, 2018. Kelly Callahan, interview by author, March 29, 2017. Terry Huang, interview by author, September 23, 2016. Terry Huang, interview by author, February 26, 2018. Matt Trowbridge, interview by author, February 3, 2017. Pennie Allen, interview by author, September 5, 2017. Pennie Allen, interview by author, April 7, 2018. Leah Frerichs, "Architecture and Design for Healthy Eating in Schools" (PhD diss., University of Nebraska, 2014). Leah Frerichs et al., "The Role of School Design in Shaping Healthy Eating–Related Attitudes, Practices, and Behaviors Among School Staff," *The Journal of School Health* 86 (2016): 11–22. Jeri Brittin, "School Design to Promote Physical Activity" (PhD diss., University of Nebraska, 2015).

69 *"When we think about movement"*: Sorensen, "Evidence-Based Design."

69 *In the long run*: Aviroop Biswas et al., "Sedentary Time and Its Association with Risk for Disease Incidence, Mortality, and Hospitalization in Adults: A Systematic Review and Meta-analysis," *Annals of Internal Medicine* 162 (2015): 123–32.

69 *In a classic study*: J. N. Morris et al., "Coronary Heart Disease and Physical Activity of Work," *Lancet* 262, no. 6795 (1953): 1053–57. J. N. Morris et al., "Coronary Heart Disease and Physical Activity of Work," *Lancet* 262, no. 6796 (1953): 1111–20.

70 *adults who sit for extended periods*: Keith M. Diaz, "Patterns of Sedentary Behavior and Mortality in U.S. Middle-Aged and Older Adults: A National Cohort Study," *Annals of Internal Medicine* 167 (2017): 465–75.

70 *The benefits of these desks*: Nipun Shrestha et al., "Workplace interventions for reducing sitting at work," *Cochrane Database of Systematic Reviews* 6 (2018): CD010912. Mark E. Benden et al., "The Impact of Stand-Biased Desks in Classrooms on Calorie Expenditure in Children," *American Journal of Public Health* 101 (2011): 1433–36. Jamilia J. Blake, Mark E. Benden, and Monica L. Wendel, "Using Stand/Sit Workstations in Classrooms: Lessons Learned from a Pilot Study in Texas," *Journal of Public Health Management and Practice* 18 (2012): 412–15.

70 *"Our big radical idea"*: Dina Sorensen, "Using Design to Promote Healthy Eating in Schools" (presentation, FitCity 10, New York, NY, May 11, 2015).

71 *Cooking classes and school gardens*: These studies are reviewed in Frerichs, "Architecture and Design," 32–38.

71 *Take a 2013 study*: Keiko Goto et al., "Do Environmental Interventions Impact Elementary School Students' Lunchtime Milk Selection?" *Applied Economic Perspectives and Policy* 35 (2013): 360–76.

71n *A clever study*: Rush University Medical Center, "Time Delays in Vending Machines Prompt Healthier Snack Choices," news release, March 30, 2017, www.rush.edu/news/press-releases/time-delays-vending-machines-prompt-healthier-snack-choices. Bradley M. Appelhans et al., "Leveraging Delay Discounting for Health: Can Time Delays Influence Food Choice?" *Appetite* 126 (2018): 16–25.

72 *Other studies have shown*: Esther Jansen, Sandra Mulkens, and Anita Jansen, "Making Fruit More Visually Appealing Increases Consumption," *Appetite* 54 (2010): 599–602. Gregory J. Privitera and Heather E. Creary, "Proximity and Visibility of Fruits and Vegetables Influence Intake in a Kitchen Setting Among College Students," *Environment and Behavior* 45 (2013): 876–86. One additional note: For years, the most prominent researcher in the field of using nudges in school cafeterias was Brian Wansink at Cornell University. But experts identified numerous errors and inconsistencies in his published work, and a number of his studies have now been retracted. In 2018, Cornell announced that a faculty committee had completed an investigation into Wansink's research practices. The committee "found that Professor Wansink committed academic misconduct in his research and scholarship, including misreporting of research data, problematic statistical techniques, failure to properly document and preserve research results, and inappropriate authorship," the provost said in a statement. (The statement is available here: Cornell University, "Statement of Cornell University Provost Michael I. Kotlikoff," news release, September 20, 2018, https://statements.cornell.edu/2018/20180920-statement-provost-michael-kotlikoff.cfm.) But that doesn't mean that the idea of nudging children into making healthier food choices is bogus. Researchers in other labs have provided evidence for this idea, and in this chapter, I have relied entirely on their work, and not on Wansink's.

72 *restaurants with glaring lights*: Nanette Stroebele and John M. De Castro, "Effect of Ambience on Food Intake and Food Choice," *Nutrition* 20 (2004): 821–38.

72 *children are more active outside*: Frerichs, "Architecture and Design."

73 *The kids loved*: Frerichs, "Architecture and Design," 113, 118–19. Allen, interview, September 5, 2017.

74 *the share of staff members*: Frerichs et al., "The Role of School Design."

74 *Other staff members*: Information on pushback and resistance from staff is from: Sorensen, interview, March 29, 2017. Allen, interview, September 5,

2017. Huang, interview, February 26, 2018. Frerichs, "Architecture and Design," 61–62, 65–67.

77 *Huang and his colleagues found*: Frerichs, "Architecture and Design," 82–101.

77 *When Huang's team used*: The findings from this study are complex and nuanced. It had been previously established that, in general, children get more sedentary as they age. And that did, in fact, happen among the children that Huang's team studied. But, through sophisticated statistical analysis and comparison to the control schools, they were able to conclude that the new school "attenuated" this increase. In other words, the data suggested that these fifth-graders were sitting less than they would have had they attended a more conventional school and, thus, that the new schools were in fact encouraging more physical activity. Information on this study and the results is from: Jeri Brittin, "Impacts of Active School Design on School-Time Sedentary Behavior and Physical Activity: A Pilot Natural Experiment," *PLOS ONE* 12 (2017): e0189236. Terry Huang, interview by author, September 23, 2016. Terry Huang, interview by author, February 26, 2018.

78 *They are reorganizing their lunch lines*: A variety of school-based active design efforts are presented in The Partnership for a Healthier New York City, *Active Design Toolkit for Schools* (2015), https://centerforactivedesign.org/dl/schools.pdf.

80 *children who have healthy habits*: The CDC has a good review of this research, with lots of fact sheets and resources: "Health and Academics," Centers for Disease Control, accessed June 5, 2019, www.cdc.gov/healthyschools/health_and_academics/index.htm.

4. THE CURE FOR THE COMMON CUBICLE

82 *These eight medical records*: My description of the Well Living Lab and its pilot study is primarily drawn from my own visit to the lab on June 20–22, 2016, as well as: Brent Bauer, interview by author, January 21, 2016. Anja Jamrozik et al., "A Novel Methodology to Realistically Monitor Office Occupant Reactions and Environmental Conditions Using a Living Lab," *Building and Environment* 130 (2018): 190–99.

82 *Background noise can impair*: Helena Jahncke et al., "Open-Plan Office Noise: Cognitive Performance and Restoration," *Journal of Environmental Psychology* 31 (2011): 373–82.

82 *Insufficient lighting*: T. L. Buchanan et al., "Illumination and Errors in Dispensing," *American Journal of Hospital Pharmacy* 48 (1991): 2137–45.

82 *Frosty and scorching air*: Li Zan, Zhiwei Lian, and Li Pan, "The Effects of

Air Temperature on Office Workers' Well-Being, Workload and Productivity-Evaluated with Subjective Ratings," *Applied Ergonomics* 42 (2010): 29–36.

83 *When the air in offices*: Joseph G. Allen et al., "Associations of Cognitive Function Scores with Carbon Dioxide, Ventilation, and Volatile Organic Compound Exposures in Office Workers: A Controlled Exposure Study of Green and Conventional Office Environments," *Environmental Health Perspectives* 124 (2016): 805–12.

83 *Thanks to our own exhalations*: W. J. Fisk, "The Ventilation Problem in Schools: Literature Review," *Indoor Air* 6 (2017): 1039–51. "Ventilation with Outdoor Air," Indoor Air Quality Scientific Findings Resource Bank, Lawrence Berkeley National Laboratory, accessed June 6, 2019, https://iaqscience.lbl.gov/topic/ventilation-outdoor-air. "Supporting Information," Indoor Air Quality Scientific Findings Resource Bank, Lawrence Berkeley National Laboratory, accessed June 6, 2019, https://iaqscience.lbl.gov/vent-info. Christopher Ingraham, "Why Crowded Meetings and Conference Rooms Make You So, So Tired," *The Washington Post*, June 6, 2019. Pan Chaoyang, "Impaired Decision Making in Conference Rooms," GIGAbase (blog), October 27, 2016, http://blog.gigabase.org/en/contents/132.

83 *in 2014, when it released*: Delos Living LLC, *The Well Building Standard* (New York: Delos Living LLC, 2014).

83 *Delos had noticed limitations*: Dana Pilai, interview by author, June 21, 2016. Dana Pilai, interview by author, August 11, 2016. Richard Macary, interview by author, June 21, 2016.

83 *A classic study*: Weiwei Liu, Weidi Zhong, and Pawel Wargocki, "Performance, Acute Health Symptoms and Physiological Responses During Exposure to High Air Temperature and Carbon Dioxide Concentration," *Building and Environment* 114 (2017): 96–105.

84 *Over the course of eighteen weeks*: Description and results of the pilot study are from my visit to the lab in 2016 as well as: Jamrozik et al., "A Novel Methodology." Brent Bauer, interview by author, October 31, 2018. Carolina Campanella and Anja Jamrozik, interview by author, June, 6, 2018. Anja Jamrozik, interview by author, October 23, 2018.

84 *the amount of data streaming in*: Alfred Anderson, interview by author, June 21, 2016.

86 *in one study, the more that researchers*: Dale Tiller et al., "Combined Effects of Noise and Temperature on Human Comfort and Performance," *ASHRAE Transactions* 116 (2010): 522–40.

86 *She worked with her colleagues*: Information about the app is from Campanella and Jamrozik, interview, June, 6, 2018. Jamrozik, interview, October 23, 2018.

86 *Daylight and window views*: Information about these two lighting studies is from: Campanella and Jamrozik, interview, June 6, 2018. Jamrozik, interview, October 23, 2018. Anja Jamrozik et al., "Access to Daylight and View in an Office Improves Cognitive Performance and Satisfaction and Reduces Eyestrain: A Controlled Crossover Study," *Building and Environment* 165 (2019): 106379.

86 *In fact, a team of German researchers*: W. U. Weitbrecht et al., "Effect of Light Color Temperature on Human Concentration and Creativity," *Fortschritte der Neurologie-Psychiatrie* 83 (2015): 344–48.

86 *An office calibrated*: Boris Kingma and Wouter van Marken Lichtenbelt, "Energy Consumption in Buildings and Female Thermal Demand," *Nature Climate Change* 5 (2015): 1054–56.

87 *in a 2019 study*: Tom Y. Chang and Agne Kajackaite, "Battle for the Thermostat: Gender and the Effect of Temperature on Cognitive Performance," *PLOS ONE* 14 (2019): e0216362.

87 *Introverts are more sensitive*: G. Belojevic, B. Jakovljevic, and V. Slepcevic, "Noise and Mental Performance: Personality Attributes and Noise Sensitivity," *Noise and Health* 6 (2003): 77–89. F. S. Morgenstern, R. J. Hodgson, and L. Law, "Work Efficiency and Personality: A Comparison of Introverted and Extraverted Subjects Exposed to Conditions of Distraction and Distortion of Stimulus in a Learning Task," *Ergonomics* 17 (1974): 211–20. Adrian Furnham and Anna Bradley, "Music While You Work: The Differential Distraction of Background Music on the Cognitive Test Performance of Introverts and Extraverts," *Applied Cognitive Psychology* 11 (1997): 445–55.

87 *In a 2011 study*: J. H. Pejtersen et al., "Sickness Absence Associated with Shared and Open-Plan Offices: A National Cross Sectional Questionnaire Survey," *Scandinavian Journal of Work, Environment and Health* 37 (2011): 376–82.

87n *In the summer of 2018*: Shane Goldmacher, "Cuomo vs. Nixon Debate? It's Already Heated (Literally)," *The New York Times*, August 28, 2018, www.nytimes.com/2018/08/28/nyregion/cuomo-nixon-debate-demands-ny.html.

88 *On the whole, surveys show*: Christhina Candido et al., "Designing Activity-Based Workspaces: Satisfaction, Productivity and Physical Activity," *Building Research and Information* 47 (2019): 275–89. Lina Engelen et al., "Is Activity-Based Working Impacting Health, Work Performance and Perceptions? A Systematic Review," *Building Research and Information* 47 (2019): 468–79.

88 *Plants can also turbocharge*: There are lots of papers and studies documenting nature's cognitive benefits. Here are a few: Jie Yin et al., "Physiological

and Cognitive Performance of Exposure to Biophilic Indoor Environment," *Building and Environment* 132 (2018): 255–62. Andrea Faber Taylor and Frances E. Kuo, "Children with Attention Deficits Concentrate Better After Walk in the Park," *Journal of Attention Disorders* 12 (2009): 402–409. Marc G. Berman, John Jonides, and Stephen Kaplan, "The Cognitive Benefits of Interacting with Nature," *Psychological Science* 19 (2008): 1207–12. Virginia I. Lohr et al., "Interior Plants May Improve Worker Productivity and Reduce Stress in a Windowless Environment," *Journal of Environmental Horticulture* 14 (1996): 97–100. Carolyn M. Tennessen and Bernadine Cimprich, "Views to Nature: Effects on Attention," *Journal of Environmental Psychology* 15 (1995): 77–85.

88 *Studies have shown*: Chih-Da Wu et al., "Linking Student Performance in Massachusetts Elementary Schools with the 'Greenness' of School Surroundings Using Remote Sensing," *PLOS ONE* 9 (2014): e108548. Dongying Li and William C. Sullivan, "Impact of Views to School Landscapes on Recovery from Stress and Mental Fatigue," *Landscape and Urban Planning* 148 (2016): 149–58. Agnes E. van den Berg et al., "Green Walls for a Restorative Classroom Environment: A Controlled Evaluation Study," *Environment and Behavior* 49 (2017): 791–813.

89 *According to Stephen*: Stephen Kaplan, "The Restorative Benefits of Nature: Toward an Integrative Framework," *Journal of Environmental Psychology* 15 (1995): 169–82.

89 *office workers who feel more control*: So Young Lee and Jay L. Brand, "Effects of Control over Office Workspace on Perceptions of the Work Environment and Work Outcomes," *Journal of Environmental Psychology* 25 (2005): 323–33. Minyoung Kwon et al., "Personal Control and Environmental User Satisfaction in Office Buildings: Results of Case Studies in the Netherlands," *Building and Environment* 149 (2019): 428–35.

89 *They're opening a second Well Living Lab*: Bauer, interview, October 31, 2018. Delos, "Delos™ to Open a Well Living Lab in Beijing, China," news release, February 22, 2017, https://delos.com/press-releases/delos-open-well -living-lab-beijing-china.

90 *The scientists hope to expand*: Bauer, interview, October 31, 2018.

90 *One of its products*: Information on the sociometric badges is from: Tanzeem Choudhury and Alex Pentland, "Sensing and Modeling Human Networks Using the Sociometer," in *Proceedings of the Seventh IEEE International Symposium on Wearable Computers* (Los Alamitos, CA: IEEE, 2003): 216–22. Taemie Kim et al., "Sociometric Badges: Using Sensor Technology to Capture New Forms of Collaboration," *Journal of Organizational Behavior* 33 (2012): 412–27. Ethan S. Bernstein and Stephen Turban, "The Impact of

the 'Open' Workspace on Human Collaboration," *Philosophical Transactions of the Royal Society B: Biological Sciences* 373 (2018): 20170239. Ben Waber, interview by author, February 27, 2019. Ethan Bernstein, interview by author, October 31, 2018.

91 *In a study of IT employees*: Lynn Wu et al., "Mining Face-to-Face Interaction Networks Using Sociometric Badges: Predicting Productivity in an IT Configuration Task," *Proceedings of the International Conference on Information Systems* (2008). Lynn Wu et al., "Mining Face-to-Face Interaction Networks Using Sociometric Badges," Humanyze, accessed May 10, 2019, www.humanyze.com/mining-face-to-face-interaction-networks-using-sociometric-badges-predicting-productivity-in-an-it-configuration-task/.

91 *Consider the case*: Information on the bank and what it learned is from: Waber, interview, February 27, 2019. "A European Bank Improves Performance Gaps Between Branches," Humanyze, accessed May 10, 2019, www.humanyze.com/case-studies-european-bank/.

92 *Over the following year*: Waber, interview, February 27, 2019. "A European Bank," Humanyze.

92 *In another study . . . jumped 10 percent*: Waber, interview, February 27, 2019. Ben Waber, "Data from the Lunchroom Could Inform the Boardroom," *The re:Work Blog*, February 24, 2016, https://rework.withgoogle.com/blog/data-from-the-lunchroom-could-inform-the-boardroom/.

92 *Research on the subject*: Some of these contradictory findings are reviewed in Bernstein and Turban, "The Impact of the 'Open' Workspace."

92 *Bernstein used the badges*: Information on the methods and results of this study comes from: Bernstein and Turban, "The Impact of the 'Open' Workspace." Bernstein, interview, October 31, 2018.

94 *by early 2019, there were more*: WeWork and HR&A Advisors, *Global Impact Report 2019* (2019), 8.

94 *but it also attracted intense scrutiny*: Lots of journalists have written about these criticisms and controversies. Here are a few of the stories: Eliot Brown, "WeWork: A $20 Billion Startup Fueled by Silicon Valley Pixie Dust," *The Wall Street Journal*, October 19, 2017, www.wsj.com/articles/wework-a-20-billion-startup-fueled-by-silicon-valley-pixie-dust-1508424483. Gaby Del Valle, "A WeWork Employee Says She Was Fired After Reporting Sexual Assault. The Company Says Her Claims Are Meritless," *Vox*, October 12, 2018, www.vox.com/the-goods/2018/10/12/17969190/wework-lawsuit-sexual-assault-harassment-retaliation. Andrew Ross Sorkin, "WeWork's Rise: How a Sublet Start-up Is Taking Over," *The New York Times*, November 13, 2018, www.nytimes.com/2018/11/13/business/dealbook/wework-office-space-real-estate.html. Matthew Yglesias, "The Controversy

Over WeWork's $47 Billion Valuation and Impending IPO, Explained," *Vox*, May 24, 2019, www.vox.com/2019/5/24/18630126/wework-valuation-ipo -business-model-we-company. Eliot Brown, "WeWork's CEO Makes Millions as Landlord to WeWork," *The Wall Street Journal*, January 16, 2019, www.wsj.com/articles/weworks-ceo-makes-millions-as-landlord-to-wework -11547640000. Reeves Wiedeman, "The I in We," *New York*, June 10, 2019, http://nymag.com/intelligencer/2019/06/wework-adam-neumann.html. Eliot Brown, "Former WeWork Executives Allege Gender, Age Discrimination," *The Wall Street Journal*, June 20, 2019, www.wsj.com/articles/former -wework-executives-allege-gender-age-discrimination-11561082242.

94 *And in the fall of 2019*: Peter Eavis and Michael J. de la Merced, "WeWork I.P.O. Is Withdrawn as Investors Grow Wary," *The New York Times*, September 30, 2019, www.nytimes.com/2019/09/30/business/wework-ipo .html.

95 *as of October 2018*: Daniel Davis, e-mail to author, October 30, 2018.

95 *The average meeting . . . or whiteboard*: Daniel Davis, interview by author, September 20, 2017. Daniel Davis, "Here's How WeWork Learns from Member Feedback," WeWork, March 24, 2016, www.wework.com/newsroom /posts/spatial-analytics.

96 *The company keeps*: Daniel Davis, interview by author, October 2, 2018. Daniel Davis, e-mail to author, October 30, 2018.

96 *Davis and his colleagues*: Davis, interview, September 20, 2017. Davis, interview, October 2, 2018. Carlo Bailey et al., "This Room Is Too Dark and the Shape Is Too Long: Quantifying Architectural Design to Predict Successful Spaces," in *Humanizing Digital Reality*, ed. Klaas De Rycke et al. (Springer Singapore, 2017), 337–48.

96 *As WeWork expanded*: Davis, interview, September 20, 2017. Davis, interview, October 2, 2018. Davis, e-mail, October 30, 2018.

97 *WeWork created software*: Information on WeWork's efforts to automate office design is from: Davis, interview, October 2, 2018. Carl Anderson et al., "Augmented Space Planning: Using Procedural Generation to Automate Desk Layouts," *International Journal of Architectural Computing* 16 (2018): 164–77. Nicole Phelan, Daniel Davis, and Carl Anderson, "Evaluating Architectural Layouts with Neural Networks," *2017 Proceedings of the Symposium on Simulation for Architecture and Urban Design* (2017): 67–73. Nicole Phelan, "Designing Offices with Machine Learning," WeWork, November 9, 2016, www.wework.com/newsroom/posts/designing-with -machine-learning. Mark Sullivan, "This Algorithm Might Design Your Next Office," WeWork, July 31, 2018, www.wework.com/newsroom/posts /this-algorithm-might-design-your-next-office.

98 *Many companies already use*: A good review of workplace surveillance and its drawbacks is: Ifeoma Ajunwa, Kate Crawford, and Jason Schultz, "Limitless Worker Surveillance," *California Law Review* 105 (2017): 735–76.

98 *Amazon has patented*: Jonathan Evan Cohn, Ultrasonic Bracelet and Receiver for Detecting Position in 2D Plane, U.S. Patent 9,881,276, filed March 28, 2016, and issued January 30, 2018.

98 *call centers have experimented*: Tom Simonite, "This Call May Be Monitored for Tone and Emotion," *Wired*, March 19, 2018, www.wired.com/story/this-call-may-be-monitored-for-tone-and-emotion/.

99 *He called it Goldilocks*: Information on Goldilocks is from: Marc Syp, interview by author, August 25, 2017. Marc Syp, interview by author, November 5, 2018.

99n *In 2015*: Arias v. Intermex Wire Transfer, LLC, et al., No. 1:15-CV-01101, complaint (Cal. Super. Ct., Bakersfield Co., May 5, 2015).

100 *Comfy, a California-based company*: Information on Comfy is from: "Product," Comfy, accessed May 10, 2019, www.comfyapp.com/product/. "What We Learned About the Workplace Experience Using Comfy @ Comfy HQ," Comfy, May 4, 2018, www.comfyapp.com/blog/what-we-learned-workplace -experience-comfy-hq/. "Keeping Employees Productive Through Thermal Comfort," Smart Buildings Center, www.smartbuildingscenter.org/sbcwp /wp-content/uploads/2015/09/SBC_CaseStudy_Comfy_lowres.pdf.

100 *In the system he designed*: "Local Warming," MIT Senseable City Lab, accessed May 10, 2019, http://senseable.mit.edu/local-warming/. "Local Warming," MIT Senseable City Lab, accessed May 10, 2019, http://senseable.mit .edu/localwarming2014/.

5. FULL SPECTRUM

103 *"I have autism"*: A video of Lindsey Eaton's speech is available at: Lindsey Eaton, "My Most Memorable Speeches," *Delve into the Power of Inclusion* (blog), March 29, 2015, https://inclusionstarters2014.blogspot.com/2015/03 /my-most-memorable-speeches.html.

104 *It was a triumphant moment*: Information about Eaton's experiences and hopes after high school is from: Lindsey and Doug Eaton, interview by author, May 11, 2018. Doug Eaton, e-mail to author, March 25, 2019. Eaton, *Delve into the Power of Inclusion* (blog), https://inclusionstarters2014 .blogspot.com/. Lindsey Eaton, "Gaining Independence," speech, SARRC Annual Community Breakfast, April 28, 2017, Phoenix, AZ, www.youtube .com/watch?v=WJFZ4jZAPcQ.

105 *Some young adults with autism*: Anne M. Roux et al., *National Autism Indicators Report: Transition into Young Adulthood* (Philadelphia: Life Course

Outcomes Research Program, A. J. Drexel Autism Institute, Drexel University, 2015).

105 *In 1961*: American Standards Association, *Making Buildings and Facilities Accessible to, and Usable by, the Physically Handicapped*, ASA A117.1-1961 (Chicago: National Society for Crippled Children and Adults, 1961).

105 *Many of these ideas*: Americans with Disabilities Act of 1990, Pub. L. 101–336.

106 *It catalyzed*: National Council on Disability, *The Impact of the Americans with Disabilities Act: Assessing the Progress toward Achieving the Goals of the ADA* (Washington, DC: National Council on Disability, 2007).

106 *What's more, designers*: National Council on Disability, *The Impact of the Americans with Disabilities Act*. Victoria Gillen, "Access for All! Neuro-architecture and Equal Enjoyment of Public Facilities," *Disability Studies Quarterly* 35 (2015). P. Hall and R. Imrie, "Architectural Practices and Disabling Design in the Built Environment," *Environment and Planning B* 26 (1999): 409–25.

106 *people with post-traumatic stress disorder*: Kunal Khanade et al., "Investigating Architectural and Space Design Considerations for Post-traumatic Stress Disorder (PTSD) Patients," *Proceedings of the Human Factors and Ergonomics Society Annual Meeting* 62 (2018): 1722–26.

107 *In 1999, the U.S. Supreme Court*: Olmstead v. L.C., 527 U.S. 581 (1999).

107 *And researchers have accumulated*: D. Felce et al., "Outcomes and Costs of Community Living: Semi-independent Living and Fully Staffed Group Homes," *American Journal of Mental Retardation* 113 (2008): 87–101. J. Howe, R. H. Horner, and J. S. Newton, "Comparison of Supported Living and Traditional Residential Services in the State of Oregon," *Mental Retardation* 36 (1998): 1–11. L. Hansson et al., "Living Situation, Subjective Quality of Life and Social Network Among Individuals with Schizophrenia Living in Community Settings," *Acta Psychiatrica Scandinavica* 106 (2002): 343–50. Roger J. Stancliffe and Sian Keane, "Outcomes and Costs of Community Living: A Matched Comparison of Group Homes and Semi-independent Living," *Journal of Intellectual and Developmental Disability* 25 (2000): 281–305. S. N. Burchard et al., "An Examination of Lifestyle and Adjustment in Three Community Residential Alternatives," *Research in Developmental Disabilities* 12 (1991): 127–42. M. L. Wehmeyer and N. Bolding, "Self-Determination Across Living and Working Environments: A Matched-Samples Study of Adults with Mental Retardation," *Mental Retardation* 37 (1999): 353–63.

108 *Medical advances and*: Molly Follette Story, James L. Mueller, and Ronald L. Mace, *The Universal Design File: Designing for People of All Ages and*

Abilities (North Carolina State University, Center for Universal Design, 1998), 6–7. "Disability and Health," World Health Organization, January 16, 2018, www.who.int/news-room/fact-sheets/detail/disability-and -health.

108 *One in ten American adults*: Catherine A. Okoro et al., "Prevalence of Disabilities and Health Care Access by Disability Status and Type Among Adults—United States, 2016," *Morbidity and Mortality Weekly Report* 67 (2018): 882–87.

108 *and a number of cognitive*: Coleen A. Boyle et al., "Trends in the Prevalence of Developmental Disabilities in US Children, 1997–2008," *Pediatrics* 127 (2011): 1034–42.

109 *For instance, when the architecture firm*: Information on the design of the new building is from: John Haymaker and Rachel Rose, interview by author, June 22, 2018. Eve Edelstein, interview by author, August 8, 2018. UC Health, "Groundbreaking at UC Health Celebrates Future Home of Neurosciences," news release, May 23, 2017, https://uchealth.com/articles /groundbreaking-at-uc-health-celebrates-future-home-of-neurosciences/. "UC Gardner Neuroscience Institute Narrated Fly-Through Video," UC Foundation, June 22, 2017, video, www.youtube.com/watch?v=qAlcw3lc2Gs. "Transforming Complex Care: Designing the New UC Gardner Neuroscience Institute," UC Health, May 24, 2017, video, www.youtube.com/watch ?v=LSnMNlXwFpM&feature=youtu.be. "First Look: University of Cincinnati Gardner Neuroscience Institute," *Healthcare Design*, October 3, 2017, www.healthcaredesignmagazine.com/projects/first-look-university -cincinnati-gardner-neuroscience-institute/.

109 *At Gallaudet*: For more information on DeafSpace, see: "DeafSpace," Gallaudet University, accessed May 13, 2019, www.gallaudet.edu/campus -design-and-planning/deafspace. Hansel Bauman, "A New Architecture for a More Livable and Sustainable World," TEDx Talks, March 6, 2015, video, www.youtube.com/watch?v=nBBdQnni9Go. Amanda Kolson Hurley, "How Gallaudet University's Architects Are Redefining Deaf Space," *Curbed*, March 2, 2016, www.curbed.com/2016/3/2/11140210/gallaudet -deafspace-washington-dc. Amanda Kolson Hurley, "Gallaudet University's Brilliant, Surprising Architecture for the Deaf," *Washingtonian*, January 13, 2016, www.washingtonian.com/2016/01/13/gallaudet-universitys -brilliant-surprising-architecture-for-the-deaf/. Sarah Holder, "How to Design a Better City for Deaf People," *CityLab*, March 4, 2019, www.citylab .com/design/2019/03/deafspace-design-disability-architecture-hard-of -hearing-dc/582613/.

110 *"It's like Yelp"*: "Sensory Inclusive App," KultureCity, accessed May 13, 2019, www.kulturecity.org/sensory-inclusive-app/. Eillie Anzilotti, "A New 'Yelp for Sensory Needs' Helps People with Autism Find Inclusive Spaces," *Fast Company*, April 2, 2018, www.fastcompany.com/40551113/a-new-yelp -for-sensory-needs-helps-people-with-autism-find-inclusive-spaces.

110 *By the time Matt*: Information about Matt Resnik is from: Denise Resnik, interview by author, October 23, 2017. Denise Resnik, interview by author, April 26, 2018. Denise Resnik, interview by author, October 26, 2018.

110 *in 1997, she cofounded*: Information about SARRC is from: Resnik, interview, October 23, 2017. Resnik, interview, April 26, 2018. Resnik, interview, October 26, 2018. "Southwest Autism Research and Resource Center," SARRC, accessed May 13, 2019, www.autismcenter.org/. "A New Vision for Life with Autism," First Place, June 8, 2018, www.firstplaceaz.org/blog/a -new-vision-for-life-with-autism/.

111 *every year, roughly fifty thousand*: Paul T. Shattuck et al., "Services for Adults with an Autism Spectrum Disorder," *Canadian Journal of Psychiatry* 57 (2012): 284–91.

111 *They published their work*: Sherry Ahrentzen and Kimberly Steele, *Advancing Full Spectrum Housing: Designing for Adults with Autism Spectrum Disorders* (Arizona State University, 2009).

113 *In the years that followed*: Information on the First Place planning process is from: Resnik, interview, October 23, 2017. Resnik, interview, April 26, 2018. Resnik, interview, October 26, 2018. Denise Resnik, e-mail to author, November 26, 2018. Mike Duffy, "Building Design Strategy and Design Goals/Guidelines" (presentation, First Place Global Leadership Institute Symposium, Phoenix, AZ, April 27, 2019).

114 *In 2012, when Kanakri*: Shireen M. Kanakri et al., "An Observational Study of Classroom Acoustical Design and Repetitive Behaviors in Children with Autism," *Environment and Behavior* 49 (2016): 847–73.

115 *It took her three years*: Shireen Kanakri, interview by author, November 15, 2017.

117 *She's been asked*: Kanakri, interview, November 15, 2017.

118 *Senior homes can help*: Gesine Marquardt, "Wayfinding for People with Dementia: A Review of the Role of Architectural Design," *HERD* 4 (2011): 75–90.

118 *Other studies suggest*: Habib Chaudhury et al., "The Influence of the Physical Environment on Residents with Dementia in Long Term Care," *The Gerontologist* 58 (2018): e325–e337.

119 *The dream is*: Information on First Place details and features is primarily

from my tour on April 26, 2018. Other sources include: Resnik, interview, October 23, 2017. Resnik, interview, April 26, 2018. Resnik, interview, October 26, 2018. Duffy, "Building Design Strategy." Mike Duffy, interview by author, April 26, 2018. "Floor Plans and Pricing Options," First Place, accessed May 13, 2019, www.firstplaceaz.org/apartments/floor-plans-pricing/. "Connected Community," First Place, accessed May 13, 2019, www.firstplaceaz.org/apartments/connected-community/. "Qualifying Criteria," First Place, accessed May 13, 2019, www.firstplaceaz.org/apartments/qualifying-criteria/.

122n *In a 2015 study*: Sharon A. Cermak et al., "Sensory Adapted Dental Environments to Enhance Oral Care for Children with Autism Spectrum Disorders: A Randomized Controlled Pilot Study," *Journal of Autism and Developmental Disorders* 45 (2015): 2876–88.

122 *The Autistic Self Advocacy Network*: Sam Crane, interview by author, December 20, 2018.

124 *In light of these concerns*: Autistic Self Advocacy Network, *ASAN's Invitational Summit on Supported Decision-Making and Transition into the Community: Summary, Conclusions, Recommendations* (ASAN, 2018), https://autisticadvocacy.org/policy/briefs/summit/.

125 *Lindsey Eaton had been*: Information about Lindsey Eaton's move and adjustment is from: Lindsey and Doug Eaton, interview by author, November 11, 2018. Lindsey Eaton, e-mail to author, March 30, 2019. Lindsey Eaton, "First Place–Phoenix, A Place to Forever Call Home," *Delve into the Power of Inclusion* (blog), October 8, 2018, https://inclusionstarters2014.blogspot.com/2018/10/first-place-phoenix-place-to-forever.html.

126 *Lauren Heimerdinger, who has autism*: Information on Lauren Heimerdinger's move and adjustment is from: Lauren Heimerdinger, interview by author, May 19, 2018. Lauren Heimerdinger, interview by author, November 29, 2018.

127 *Her son, Matt*: Information on Matt Resnik's transition is from: Resnik, interview, October 26, 2018. Denise Resnik, "New Life Chapters Tie Together Bingo, Besties and the Beatles," First Place, December 13, 2018, www.firstplaceaz.org/blog/new-life-chapters-tie-together-bingo-besties-and-the-beatles/.

128 *In the long run*: Resnik, interview, October 26, 2018.

6. JAILBREAKERS

129 *Davis had a turbulent childhood*: Information on Davis's childhood and his early criminal history is from Anthony Davis, letter to author, April 22, 2018. Anthony Davis, letter to author, September 16, 2018.

130 *Years earlier, Davis told me*: This account is based on Davis's own description of the crime, which he relayed to me in a phone call: Anthony Davis, interview by author, July 3, 2018.

130 *and Davis was sentenced*: My descriptions of Davis's experiences in prison in general, and solitary in particular, are based on his accounts, which he relayed to me in a series of phone calls and letters. Information on Davis's first trip to solitary is from: Anthony Davis, interview by author, July 2, 2018. Davis, letter, April 22, 2018.

131 *There are more people*: E. Fuller Torrey et al., *More Mentally Ill Persons Are in Jails and Prisons Than Hospitals: A Survey of the States* (Treatment Advocacy Center and National Sheriffs' Association, 2010).

132 *with more than 2 million*: Danielle Kaeble and Mary Cowhig, *Correctional Populations in the United States, 2016* (U.S. Department of Justice, Bureau of Justice Statistics, 2018), www.bjs.gov/content/pub/pdf/cpus16.pdf.

132 *For most of history*: Information on the early history of jails and prisons, and their design, is largely from Richard E. Wener, *The Environmental Psychology of Prisons and Jails*, reprint ed. (Cambridge, UK: Cambridge University Press, 2014), 12–42.

132n *Globally, there are*: Penal Reform International, *Global Prison Trends 2015* (London: PRI, 2015).

133 *"short and massive"*: This quote is from Wener, *Environmental Psychology*, 33–34.

133 *And, at Eastern State*: Information on the history, design, and psychological effects of Eastern State is from: Wener, *Environmental Psychology*, 23–29, 38. Peter Scharff Smith, "The Effects of Solitary Confinement on Prison Inmates: A Brief History and Review of the Literature," *Crime and Justice* 34 (2006): 441–528. Stuart Grassian, "Psychiatric Effects of Solitary Confinement," *Washington University Journal of Law and Policy* 22 (2006): 325–83.

134 *In his 1847 report*: Francis C. Gray, *Prison Discipline in America* (Boston: Little, Brown, 1847), 181.

134 *many American localities*: Information about the Auburn model is from: Wener, *Environmental Psychology*, 29–33, 38–40. Gordon S. Bates, *The Connecticut Prison Association and the Search for Reformatory Justice* (Middletown, CT: Wesleyan University Press, 2017), 63–66.

135 *On the morning of October 22, 1983 . . . stabbing another guard*: Information about these murders is from: United States v. Fountain, 768 F.2d 790 (7000 Cir. 1985). "Merle E. Clutts," Federal Bureau of Prisons, accessed May 14, 2019, www.bop.gov/about/history/hero_clutts.jsp?i=17. "Robert L. Hoffman," Federal Bureau of Prisons, accessed May 14, 2019, www.bop.gov/about/history/hero_hoffmann.jsp?i=18.

135 *At the time*: The history of Marion is reviewed in: David A. Ward and Thomas G. Werlich, "Alcatraz and Marion: Evaluating Super-maximum Custody," *Punishment and Society* 5 (2003): 53–75. Stephen C. Richards, "USP Marion: The First Federal Supermax," *The Prison Journal* 88 (2008): 6–22.

135 *Marion put all its inmates*: Information on the lockdown, and the conditions that resulted, is from: Ward and Werlich, "Alcatraz and Marion." Richards, "USP Marion." Fay Dowker and Glenn Good, *From Alcatraz to Marion to Florence: Control Unit Prisons in the United States* (Committee to End the Marion Lockdown, 1972). Dennis Cunningham and Jan Susler, *A Public Report About Violent Mass Assault Against Prisoners and Continuing Illegal Punishment and Torture of the Prison Population at the U.S. Penitentiary at Marion, Illinois* (Marion Prisoners' Rights Project, 1983). *Capsule Summary of the Marion Lockdown* (Committee to End the Marion Lockdown, 1983). E. R. Shipp, "Killings Tighten Rule at Tough Prison," *The New York Times*, January 20, 1984, www.nytimes.com/1984/01/20/us/killings-tighten-rule-at-tough-prison.html. *"An Uneasy Calm...": The U.S. Penitentiary at Marion* (John Howard Association, 1986).

136 *"an uneasy calm prevails"*: *"An Uneasy Calm,"* 16.

136 *The supermax building boom*: The proliferation of supermax prisons and segregation units in America is detailed in: Ward and Werlich, "Alcatraz and Marion." Richards, "USP Marion." Russ Immarigeon, "The Marionization of American Prisons," *National Prison Project Journal* 7 (1992): 1–5. Smith, "The Effects of Solitary Confinement," 443.

136 *Solitary confinement became*: Angela Browne, Alissa Cambier, and Suzanne Agha, "Prisons Within Prisons: The Use of Segregation in the United States," *Federal Sentencing Reporter* 24 (2011): 46–49.

136 *experts estimate that*: Sarah Baumgartel et al., "Time-in-Cell: The ASCA-Liman 2014 National Survey of Administrative Segregation in Prison," Yale Law School, Public Law Research Paper No. 552 (2015), https://papers.ssrn.com/sol3/papers.cfm?abstract_id=2655627.

136n *in 2016*, The Guardian *reported*: Aviva Stahl, "Concern over 'Political' Use of Solitary Confinement in European Prisons," *The Guardian*, May 2, 2016, www.theguardian.com/world/2016/may/02/solitary-confinement-european-prisons-terror-threat.

137 *After murdering*: Sam Roberts, "Thomas Silverstein, Killer and Most Isolated Inmate, Dies at 67," *The New York Times*, May 21, 2019, www.nytimes.com/2019/05/21/obituaries/thomas-silverstein-dead.html.

137 *For Anthony Davis*: Descriptions of Davis's experiences in solitary are from: Anthony Davis, letter to author, April 1, 2018. Davis, letter, April 22, 2018. Davis, interview, July 2, 2018. Davis, interview, July 3, 2018.

138 *When Craig Haney*: Craig Haney, "Mental Health Issues in Long-Term Solitary and 'Supermax' Confinement," *Crime and Delinquency* 49 (2003): 124–56.

138 *the symptoms occur*: Smith, "The Effects of Solitary Confinement."

139 *In a 2014 study*: Fatos Kaba, "Solitary Confinement and Risk of Self-Harm Among Jail Inmates," *American Journal of Public Health* 104 (2014): 442–47.

139 *In the 1950s*: Woodburn Heron, "The Pathology of Boredom," *Scientific American*, January 1957, 52–56.

140 *One inmate I corresponded with*: Descriptions of Harris's experiences in solitary are from: Francis Harris, e-mail to author, May 15, 2018. Francis Harris, e-mail to author, July 30, 2018. Francis Harris, e-mail to author, August 13, 2018. Francis Harris, e-mail to author, October 2, 2018. Francis Harris, e-mail to author, November 9, 2018. Francis Harris, e-mail to author, March 23, 2019.

140 *Ellard has demonstrated*: Information on Ellard's research is from: Colin Ellard, "A New Agenda for Urban Psychology: Out of the Laboratory and Onto the Streets," *Journal of Urban Design and Mental Health* 2 (2017), www.urbandesignmentalhealth.com/journal2-ellard.html. Colin Ellard, interview by author, March 6, 2017.

141 *The neuroscientist James Danckert*: Colleen Merrifield and James Danckert, "Characterizing the Psychophysiological Signature of Boredom," *Experimental Brain Research* 232 (2014): 481–91.

141 *Loneliness and social isolation*: There are lots of studies on the effects of loneliness and isolation in humans and other animals. Here are a few good reviews: Julianne Holt-Lunstad et al., "Loneliness and Social Isolation as Risk Factors for Mortality: A Meta-analytic Review," *Perspectives of Psychological Science* 10 (2015): 227–37. John T. Cacioppo et al., "The Neuroendocrinology of Social Isolation," *Annual Review of Psychology* 66 (2015): 733–67. Louise C. Hawkley and John T. Cacioppo, "Loneliness Matters: A Theoretical and Empirical Review of Consequences and Mechanisms," *Annals of Behavioral Medicine* 40 (2010): 218–27.

142 *"The charming, funny"*: Anthony Davis, "Voices from Solitary: At War with My Own Self," Solitary Watch, September 30, 2014, https://solitarywatch.org/2014/09/30/voices-from-solitary-at-war-with-my-own-self/.

143 *The United States comprises*: According to data published by the World Prison Population List. As of September 2018, there were 10.74 million prisoners worldwide, with more than 2.1 million of them held in the United States. Roy Walmsley, *World Prison Population*, 12th ed. (London: Institute for Criminal Policy Research, 2018), 2.

143 *The archetypical example*: Halden Prison and the Norwegian Correctional Service recently published a magazine about the prison and the correctional philosophy behind it. It is: *Halden Prison: Punishment That Works— Change That Lasts!* (Halden fengsel: 2019), https://issuu.com/omdocs/docs /magasin_halden_prison_issu. A number of journalists have also written about Halden. Some good introductions are Jessica Benko, "The Radical Humaneness of Norway's Halden Prison," *The New York Times Magazine*, March 26, 2015, www.nytimes.com/2015/03/29/magazine/the-radical -humaneness-of-norways-halden-prison.html. Amelia Gentleman, "Inside Halden, the Most Humane Prison in the World," *The Guardian*, May 18, 2012, www.theguardian.com/society/2012/may/18/halden-most-humane -prison-in-world. William Lee Adams, "Sentenced to Serving the Good Life in Norway," *Time*, July 12, 2010, http://content.time.com/time/magazine /article/0,9171,2000920,00.html.

145 *The HMC-KMD team*: Information on the design and operations of Las Colinas is from my own visit on March 6, 2018, as well as: James Krueger, interview by author, March 6, 2018. James Krueger, Vern Almon, and Steve Carter, interview by author, February 28, 2017. Steve Carter, interview by author, April 13, 2018. Christine Brown-Taylor, interview by author, April 12, 2018. Christine Brown-Taylor, e-mail to author, September 4, 2018.

147 *Several studies suggest*: Nalini M. Nadkarni et al., "Impacts of Nature Imagery on People in Severely Nature-Deprived Environments," *Frontiers in Ecology and the Environment* 15 (2017): 395–403. Jay Farbstein, Melissa Farling, and Richard Wener, *Effects of a Simulated Nature View on Cognitive and Psycho-physiological Responses of Correctional Officers in a Jail Intake Area* (2009).

147 *The Federal Bureau of Prisons*: Information on the history and benefits of direct supervision is from: Wener, *Environmental Psychology*, 46–107. Richard Wener, interview by author, June 21, 2017.

148n *A 2014 study*: Karin A. Beijersbergen et al., "A Social Building? Prison Architecture and Staff-Prisoner Relationships," *Crime and Delinquency* 62 (2014): 843–74.

150 *In 2015, CGL*: Information on the post-occupancy evaluation is from an unpublished draft copy of the POE that was provided to me as well as Carter, interview, April 13, 2018.

150 *after the new facility opened*: Brown-Taylor, interview by author, April 12, 2018.

152 *a substantial subset of Americans*: In a 2016 survey, 18 percent of Americans said that the purpose of jails was to punish. RTI International, "Majority

of Americans Believe Role of Jails Should Not Be to Punish," news release, March 17, 2017, www.rti.org/news/majority-americans-believe-role-jails -should-not-be-punish.

152 *"The mayor wants jails"*: Michael Gartland and Laura Italiano, "The Mayor Wants Jails That Feel Like a Retreat in Tulum," *New York Post*, June 22, 2017, https://nypost.com/2017/06/22/nyc-basically-wants-to-replace-rikers -with-a-daycare/.

154 *In the United States, ten thousand people*: "Prisoners and Prisoner Re-entry," U.S. Department of Justice, accessed May 15, 2017, www.justice.gov/archive /fbci/progmenu_reentry.html.

154 *Psychiatric hospitals are increasingly*: Francis Pitts, interview by author, July 24, 2018. Francis Pitts, "Design of Mental Health Facilities: A Primer" (pre-sentation, AIA Conference on Architecture, New York, NY, June 22, 2018).

155 *In a 2012 study*: Lorraine E. Maxwell and Suzanne L. Schechtman, "The Role of Objective and Perceived School Building Quality in Student Aca-demic Outcomes and Self-Perception," *Children, Youth and Environments* 22 (2012): 23–51.

155 *People who live in litter-strewn*: *Assembly Civic Design Guidelines* (Center for Active Design, 2018), 31.

155 *including bolstering mental health*: Eugenia C. South et al., "Effect of Greening Vacant Land on Mental Health of Community-Dwelling Adults: A Cluster Randomized Trial," *JAMA Network Open* 1 (2018): e180298.

155 *people who live on blocks: Assembly Civic Design Guidelines*, 40.

155 *Sprucing up vacant lots*: Charles C. Branas et al., "A Difference-in-Differences Analysis of Health, Safety, and Greening Vacant Urban Space," *American Journal of Epidemiology* 174 (2011): 1296–1306. Charles C. Branas et al., "Citywide Cluster Randomized Trial to Restore Blighted Vacant Land and Its Effects on Violence, Crime, and Fear," *PNAS* 115 (2018): 2946–51.

156 *Bennett sees clear parallels*: Kevin Bennett, Tyler Gualtieri, and Becky Kaz-mierczyk, "Undoing Solitary Urban Design: A Review of Risk Factors and Mental Health Outcomes Associated with Living in Social Isolation," *Journal of Urban Design and Mental Health* 4 (2018), www.urbandesignmentalhealth .com/journal-4---solitary-urban-design.html.

156 *In a 2012 survey*: The survey results are detailed in: *Connections and Engage-ment: A Survey of Metro Vancouver* (Vancouver Foundation, 2012). *Con-nections and Engagement Closer Look: The Effect of Apartment Living on Neighbourliness* (Vancouver Foundation, 2012), www.vancouverfoundation .ca/about-us/publications/connections-and-engagement-reports /connections-engagement-closer-look-effect.

156 *Studies of lounges*: Robert Sommer and Hugo Ross, "Social Interaction on a Geriatrics Ward," *International Journal of Social Psychiatry* 4 (1958): 128–33. Robert F. Peterson et al., "The Effects of Furniture Arrangement on the Behavior of Geriatric Patients," *Behavior Therapy* 8 (1977): 464–67.

157 *Researchers have also found*: Daniel A. Cox and Ryan Streeter, *The Importance of Place: Neighborhood Amenities as a Source of Social Connection and Trust* (American Enterprise Institute: 2019), www.aei.org/wp-content/uploads/2019/05/The-Importance-of-Place.pdf.

157 *Finally, promoting urban well-being*: Andrew J. Hoisington et al., "Ten Questions Concerning the Built Environment and Mental Health," *Building and Environment* 155 (2019): 58–69.

157 *in one large, longitudinal study*: Rebecca Bentley et al., "Association Between Housing Affordability and Mental Health: A Longitudinal Analysis of a Nationally Representative Household Survey in Australia," *American Journal of Epidemiology* 174 (2011): 753–60.

7. IF THESE WALLS COULD TALK, LISTEN, AND RECORD

160 *By 2023, more than half*: "Smart Home: United States," Statista, accessed May 16, 2019, www.statista.com/outlook/279/109/smart-home/united-states. "Smart Home: Worldwide," Statista, accessed May 16, 2019, www.statista.com/outlook/279/100/smart-home/worldwide.

160 *Google has patented*: Brian Derek DeBusschere and Jeffrey L. Rogers, Assessing Cardiovascular Function Using an Optical Sensor, U.S. Patent 9,848,780, filed April 8, 2015, and issued December 26, 2017.

160 *Amazon has patented*: Huafeng Jin and Shuo Wang, "Voice-Based Determination of Physical and Emotional Characteristics of Users," U.S. Patent 10,096,319, filed March 13, 2017, and issued October 9, 2018.

161 *There are more than 700 million*: *World Population Prospects 2019: Highlights* (New York: United Nations, 2019).

161 *Transferring elderly men*: Pamela S. Manion and Marilyn J. Rantz, "Relocation Stress Syndrome: A Comprehensive Plan for Long-Term Care Admissions," *Geriatric Nursing* 16 (1995): 108–12. Sonya Brownie, Louise Horstmanshof, and Rob Garbutt, "Factors That Impact Residents' Transition and Psychological Adjustment to Long-Term Aged Care: A Systematic Literature Review," *International Journal of Nursing Studies* 51 (2014): 1654–66.

161 *after the staff moved*: Marilyn Rantz and Kathleen Egan, "Reducing Death from Translocation Syndrome," *American Journal of Nursing* 87 (1987): 1351–53.

162 *Seniors aren't too keen*: Marilyn J. Rantz et al., "TigerPlace, A State-

Academic-Private Project to Revolutionize Traditional Long-Term Care," *Journal of Housing for the Elderly* 22 (2008): 66–85. Marilyn J. Rantz et al., "A Technology and Nursing Collaboration to Help Older Adults Age in Place," *Nursing Outlook* 53 (2005): 40–45. Joanne Binette and Kerri Vasold, *2018 Home and Community Preferences: A National Survey of Adults Age 18-Plus* (Washington, DC: AARP Research, 2018). Linda Barrett, *Home and Community Preferences of the 45+ Population* (Washington, DC: AARP Research, 2015).

162 *In 1996, Rantz*: Information on the origins of TigerPlace is from: Rantz et al., "TigerPlace." Karen Marek and Marilyn Rantz, "Aging in Place: A New Model for Long-Term Care," *Nursing Administration Quarterly* 24 (2000): 1–11. Marilyn Rantz, interview by author, May 15, 2018. Marilyn Rantz, interview by author, February 13, 2019.

162 *More than one in four*: "Older Adult Falls," *Centers for Disease Control and Prevention*, accessed May 16, 2017, www.cdc.gov/homeandrecreationalsafety/falls/adultfalls.html. Jane Fleming and Carol Brayne, "Inability to Get Up After Falling, Subsequent Time on Floor, and Summoning Help: Prospective Cohort Study in People over 90," *BMJ* 337 (2008): a2227.

163 *In a series of focus groups*: G. Demiris et al., "Older Adults' Attitudes Towards and Perceptions of 'Smarthome' Technologies: A Pilot Study," *Medical Informatics and the Internet in Medicine* 29 (2004): 87–94.

163 *Working with several colleagues*: Information on the development of the fall detection system is from: Marjorie Skubic, interview by author, May 11, 2018. Marjorie Skubic, interview by author, June 14, 2018. Marjorie Skubic, interview by author, January 9, 2019. Erik Stone and Marjorie Skubic, "Fall Detection in Homes of Older Adults Using the Microsoft Kinect," *IEEE Journal of Biomedical and Health Informatics* 19 (2015): 290–301. Marilyn J. Rantz et al., "Falls, Technology, and Stunt Actors: New Approaches to Fall Detection and Fall Risk Assessment," *Journal of Nursing Care Quality* 23 (2008): 195–201.

164 *The sprawling, one-story*: Information about TigerPlace is from my own visit on June 14, 2018, plus: Rantz, interview, May 15, 2018. Rantz, interview, February 13, 2019. Kari Lane, interview by author, May 16, 2018. Kari Lane, interview by author, June 14, 2018. Rantz et al., "TigerPlace."

164 *The residents range . . . a dozen medications*: Marilyn J. Rantz et al., "Sensor Technology to Support Aging in Place," *Journal of the American Medical Directors Association* 14 (2013): 386–91.

165 *the average length of stay*: Marilyn Rantz et al., "The Continued Success of Registered Nurse Care Coordination in a State Evaluation of Aging in Place in Senior Housing," *Nursing Outlook* 62 (2014): 237–46.

165 *Skubic's team launched*: Information on the methods and results of this study comes from: Erik Stone and Marjorie Skubic, "Testing Real-Time In-Home Fall Alerts with Embedded Depth Video Hyperlink," in *Smart Homes and Health Telematics: 12th International Conference, ICOST 2014*, ed. Cathy Bodine et al. (Switzerland: Springer International, 2015), 41–48.

165 *The system proved popular*: Marilyn Rantz et al., "Automated In-Home Fall Risk Assessment and Detection Sensor System for Elders," *Gerontologist* 55 (2015): S78–S87.

166 *When they tested it*: Lorraine J. Phillips et al., "Using Embedded Sensors in Independent Living to Predict Gait Changes and Falls," *Western Journal of Nursing Research* 39 (2017): 78–94.

166 *Skubic's team also*: Information about the illness detection system and its successes is from: Skubic, interview, May 11, 2018. Skubic, interview, June 14, 2018. Skubic, interview, January 9, 2019. Lane, interview, May 16, 2018. Lane, interview, June 14, 2018. Rantz, interview, May 15, 2018. Phillips et al., "Using Embedded Sensors." Marilyn J. Rantz, "Automated Technology to Speed Recognition of Signs of Illness in Older Adults," *Journal of Gerontological Nursing* 38 (2012): 18–23. Marilyn Rantz et al., "Using Technology to Enhance Aging in Place," in *Smart Homes and Health Telematics: 6th International Conference, ICOST 2008*, ed. Sumi Helal et al. (Germany: Springer, 2008), 169–76. Marilyn Rantz et al., "Improving Nurse Care Coordination with Technology," *Computers, Informatics, Nursing* 28 (2010): 325–32. Marilyn Rantz et al., "Enhanced Registered Nurse Care Coordination with Sensor Technology: Impact on Length of Stay and Cost in Aging in Place Housing," *Nursing Outlook* 63 (2015): 650–55.

167 *In a 2015 study*: Marilyn Rantz et al., "Enhanced Registered Nurse Care Coordination."

168 *Foresite has developed . . . multiple states*: George Chronis, interview by author, May 24, 2018. George Chronis, e-mail to author, January 18, 2019.

168 *a lesson that Skubic learned*: Information on Skubic's father and his illness is from Skubic, interview, May 11, 2018.

170 *While many seniors are*: Information about seniors' attitudes toward the technology is from: Lane, interview, June 14, 2018. G. Demiris et al., "Older Adults' Attitudes Towards and Perceptions of 'Smarthome' Technologies: A Pilot Study," *Medical Informatics and the Internet in Medicine* 29 (2004): 87–94. George Demiris et al., "Senior Residents' Perceived Need of and Preferences for 'Smart Home' Sensor Technologies," *International Journal of Technology in Health Care* 24 (2008): 120–24. Karen Courtney et al., "Needing Smart Home Technologies: The Perspectives of Older Adults in Continuing Care Retirement Communities," *Journal of Innovation in*

Health Informatics 16 (2008): 195–201. E. Robinson et al., "Creating a Tailored, In-Home, Sensor System to Facilitate Healthy Aging: The Consumer Perspective," *Innovation in Aging* 2 (2018): 912.

171 *A group of researchers*: Toshiyo Tamura et al., "E-Healthcare at an Experimental Welfare Techno House in Japan," *The Open Medical Informatics Journal* 1 (2007): 1–7.

171 *A team of MIT engineers*: Fadel Adib et al., "Smart Homes That Monitor Breathing and Heart Rate," *Proceedings of the 33rd Annual ACM Conference on Human Factors in Computing Systems* (2015): 837–46.

171 *There are motion-detector-based systems*: Anthony P. Glascock and David M. Kutzik, "An Evidentiary Study of the Uses of Automated Behavioral Monitoring," in *21st International Conference on Advanced Information Networking and Applications Workshops* (Los Alamitos, CA: IEEE, 2007): 858–62. "The Impact of Behavioral Monitoring Technology on the Provision of Health Care in the Home," *Journal of Universal Computer Science* 12 (2006): 59–79.

171 *intelligent assistants that remind*: Eren Demir et al., "Smart Home Assistant for Ambient Assisted Living of Elderly People with Dementia," *Procedia Computer Science* 113 (2017): 609–14.

171 *There are even robots*: Minh Pham et al., "Delivering Home Healthcare Through a Cloud-Based Smart Home Environment (CoSHE)," *Future Generation Computer Systems* 81 (2018): 129–40. Ha Manh Do et al., "RiSH: A Robot-Integrated Smart Home for Elderly Care," *Robotics and Autonomous Systems* 101 (2018): 74–92.

171 *Elder care robots*: Roger Bemelmans et al., "Socially Assistive Robots in Elderly Care: A Systematic Review into Effects and Effectiveness," *Journal of American Medical Directors Association* 13 (2012): 114–210.e1. Elaine Mordoch et al., "Use of Social Commitment Robots in the Care of Elderly People with Dementia: A Literature Review," *Maturitas* 74 (2013): 14–20. Malcolm Foster, "Aging Japan: Robots May Have Role in Future of Elder Care," Reuters, March 27, 2018, www.reuters.com/article/us-japan-ageing -robots-widerimage/aging-japan-robots-may-have-role-in-future-of-elder -care-idUSKBN1H33AB. Daniel Hurst, "Japan Lays Groundwork for Boom in Robot Carers," *The Guardian*, February 5, 2018, www.theguardian.com /world/2018/feb/06/japan-robots-will-care-for-80-of-elderly-by-2020.

171 *The U.S. Department of Veterans Affairs*: Kristina M. Martinez et al., "VA SmartHome for Veterans with TBI: Implementation in Community Settings," in *Smart Homes and Health Telematics: 12th International Conference, ICOST 2014*, ed. Cathy Bodine et al. (Switzerland: Springer International, 2015), 110–18.

172 *In the spring of 2018*: Information on Colistra's ideas and systems are from my visit to Lawrence, Kansas, on March 27, 2018, as well as: Joe Colistra, interview by author, September 14, 2017. Joe Colistra, interview by author, March 27, 2018. Joe Colistra, "The Evolving Architecture of Smart Cities," in *2018 IEEE International Smart Cities Conference* (Los Alamitos, CA: IEEE, 2018).

174 *In October 2018*: Sapna Maheshwari, "This Thermometer Tells Your Temperature, Then Tells Firms Where to Advertise," *The New York Times*, October 23, 2018, www.nytimes.com/2018/10/23/business/media/fever -advertisements-medicine-clorox.html.

174 *A month later, ProPublica*: Marshall Allen, "You Snooze, You Lose: Insurers Make the Old Adage Literally True," *ProPublica*, November 21, 2018, www .propublica.org/article/you-snooze-you-lose-insurers-make-the-old-adage -literally-true.

174n *One example of just how personal*: Kashmir Hill, "This Sex Toy Tells the Manufacturer Every Time You Use It," *Splinter*, August 9, 2016, https:// splinternews.com/this-sex-toy-tells-the-manufacturer-every-time-you -use-1793861000. Camila Domonoske, "Vibrator Maker to Pay Millions Over Claims It Secretly Tracked Use," NPR, March 14, 2017, www.npr .org/sections/thetwo-way/2017/03/14/520123490/vibrator-maker-to-pay -millions-over-claims-it-secretly-tracked-use.

175 *Some digital health companies*: Kit Huckvale, John Torous, and Mark E. Larsen, "Assessment of the Data Sharing and Privacy Practices of Smartphone Apps for Depression and Smoking Cessation," *JAMA Network Open* 2 (2019): e192542.

175n *In one large, randomized*: Kevin G. Volpp et al., "Effect of Electronic Reminders, Financial Incentives, and Social Support on Outcomes After Myocardial Infarction: The HeartStrong Randomized Clinical Trial," *JAMA Internal Medicine* 177 (2017): 1093–1101.

176 *Some facial recognition software*: Joy Buolamwini and Timnit Gebru, "Gender Shades: Intersectional Accuracy Disparities in Commercial Gender Classification," *Proceedings of Machine Learning Research* 81 (2018): 1–15.

176 *Skubic has also noticed*: Skubic, interview, January 9, 2019.

176 *strong social networks are key*: Yeol Choi, Yeon-Hwa Kwon, and Jeongseob Kim, "The Effect of the Social Networks of the Elderly on Housing Choice in Korea," *Habitat International* 74 (2018): 1–8.

177 *He's developing a system*: Juan Carlos Augusto, interview by author, December 14, 2018.

177 *Augusto has proposed*: Juan Carlos Augusto, interview by author, September 26, 2017. Simon Jones, Sukhvinder Hara, and Juan Carlos Augusto, "eFRIEND: an Ethical Framework for Intelligent Environments Development," *Ethics and Information Technology* 17 (2015): 11–25.

178 *About a decade ago, Schnädelbach*: Information about ExoBuilding is from my visit to Schnädelbach's lab on December 1, 2017, as well as Holger Schnädelbach, interview by author, November 21, 2016. Holger Schnädelbach, interview by author, December 1, 2017. Holger Schnädelbach, Kevin Glover, and Ainojie Alexander Irune, "ExoBuilding: Breathing Life into Architecture," in *NordiCHI 2010: Proceedings of the 6th Nordic Conference on Human-Computer Interaction* (New York: ACM Press, 2010), 442–51. Holger Schnädelbach et al., "ExoBuilding: Physiologically Driven Adaptive Architecture," *ACM Transactions on Computer-Human Interaction* 19 (2012). Stuart Moran et al., "ExoPranayama: A Biofeedback-Driven Actuated Environment for Supporting Yoga Breathing Practices," *Personal and Ubiquitous Computing* 20 (2016): 261–75.

179 *In a 2016 paper*: Antonio Fernández-Caballero et al., "Smart Environment Architecture for Emotion Detection and Regulation," *Journal of Biomedical Informatics* 64 (2016): 55–73.

8. HOPE FLOATS

182 *between 1995 and 2015*: Centre for Research on the Epidemiology of Disasters (CRED) and United Nations Office for Disaster Risk Reduction (UNDRR), *The Human Cost of Weather Related Disasters: 1995–2015* (Brussels: CRED and UNDRR, 2015), 13.

182 *Although the precise effects*: The many consequences of climate change are laid out in: D. R. Reidmiller et al., eds., *Impacts, Risks, and Adaptation in the United States: Fourth National Climate Assessment*, vol. II (Washington, DC: U.S. Global Change Research Program, 2018). V. Masson-Delmotte et al., eds., *Global Warming of 1.5°C* (Intergovernmental Panel on Climate Change, 2018). U.S. Environmental Protection Agency, *Climate Change Indicators in the United States, 2016*, 4th ed. (EPA, 2016).

182 *Hurricane Harvey dumped*: Eric S. Blake and David A. Zelinsky, *National Hurricane Center Tropical Cyclone Report: Hurricane Harvey*, report no. AL092017 (National Oceanic and Atmospheric Administration and National Weather Service, May 9, 2018), www.nhc.noaa.gov/data/tcr/AL092017_Harvey.pdf.

183 *Monsoons killed*: UNICEF, "16 Million Children Affected by Massive Flooding in South Asia, with Millions More at Risk," news release, September 2, 2017, www.unicef.org/infobycountry/media_100719.html.

183 *Heavy rains in Sierra Leone*: United Nations Office for the Coordination of Humanitarian Affairs, *West and Central Africa: 2017 Flood Impact* (October 18, 2017), https://reliefweb.int/report/niger/west-and-central-africa-2017 -flood-impact-18-oct-2017.

183 *Record-breaking heat waves*: Sarah F. Kew et al., "The Exceptional Summer Heat Wave in Southern Europe 2017," *Bulletin of the American Meteorological Society* 100 (2019): S49–S53. Chunlüe Zhou et al., "Attribution of a Record-Breaking Heatwave Event in Summer 2017 over the Yangtze River Delta," *Bulletin of the American Meteorological Society* 100 (2019): S97–S103.

183 *fires consumed California*: "California Statewide Fire Summary, Monday, October 30, 2017," CAL FIRE, http://calfire.ca.gov/communications /communications_StatewideFireSummary.

183 *As many as 40 percent*: Emily Goldmann and Sandro Galea, "Mental Health Consequences of Disasters," *Annual Review of Public Health* 35 (2014): 169–83.

183 *Elizabeth English has dedicated*: Information about English's background and the origins of the Buoyant Foundation Project is from: Elizabeth English, interview by author, August 2, 2017. Elizabeth English, interview by author, August 15, 2017. Elizabeth English, interview by author, August 8, 2018.

184 *Eighty percent of the city . . . never came back*: Richard D. Knabb, Jamie R. Rhome, and Daniel P. Brown, *Tropical Cyclone Report: Hurricane Katrina*, rev. ed. (National Oceanic and Atmospheric Administration, September 14, 2011), www.nhc.noaa.gov/data/tcr/AL122005_Katrina.pdf. U.S. Department of Housing and Urban Development, *Current Housing Unit Damage Estimates: Hurricanes Katrina, Rita, and Wilma* (2006), www.huduser.gov /publications/pdf/GulfCoast_Hsngdmgest.pdf. Elizabeth Fussell, "The Long Term Recovery of New Orleans' Population After Hurricane Katrina," *The American Behavioral Scientist* 59 (2015): 1231–45. Brian Wolshon, "Evacuation Planning and Engineering for Hurricane Katrina," *The Bridge* 36 (2006): 27–34. Jean-Marc Zaninetti and Craig E. Colten, "Shrinking New Orleans: Post-Katrina Population Adjustments," *Urban Geography* 33 (2012): 675–99.

185 *including the residents of Belén*: Natasha Klink, "Amphibious Amazon: Traditional Approaches to Amphibiation in Peru" (presentation, ICAADE 2017, Waterloo, Canada, June 26, 2017).

185 *The houses they built . . . their original positions*: Elizabeth Victoria Fenuta, "Amphibious Architectures: The Buoyant Foundation Project in Post-Katrina New Orleans (master's thesis, University of Waterloo, 2010), 167–91.

185 *One of the students told English*: Information on Raccourci Old River is from: Fenuta, "Amphibious Architectures," 160–66. English, interview, August 2, 2017. "Forty Years of Amphibious Housing in Old River Landing, Louisiana" (presentation, ICAADE 2017, Waterloo, Canada, June 26, 2017).

186 *"I got tired of that"*: Buddy Blalock, "Forty Years of Amphibious Housing in Old River Landing, Louisiana" (remote presentation, ICAADE 2017, Waterloo, Canada, June 26, 2017).

186 *"I've been out here"*: Blalock, "Forty Years."

187 *English thought that she could make it amphibious*: Information about English's solution and its advantages is from: English, interview, August 2, 2017. English, interview, August 15, 2017. Fenuta, "Amphibious Architectures." Elizabeth English, Natasha Klink, and Scott Turner, "Thriving with Water: Developments in Amphibious Architecture in North America," *E3S Web of Conferences* 7 (2016): 13009.

188 *The hazards of climate change*: *World Economic and Social Survey 2016: Climate Change Resilience—An Opportunity for Reducing Inequalities* (United Nations Economic and Social Council, 2016).

188 *In Miami*: Jesse M. Keenan, Thomas Hill, and Anurag Gumber, "Climate Gentrification: From Theory to Empiricism in Miami-Dade County, Florida," *Environmental Research Letters* 13 (2018): 054001.

188 *Hurricane Katrina laid . . . than poor ones*: *World Economic and Social Survey*, 5, 36, 43, 88. Craig E. Colten, "Vulnerability and Place: Flat Land and Uneven Risk in New Orleans," *American Anthropologist* 108 (2006): 731–34. Michel Masozera, Melissa Bailey, and Charles Kerchner, "Distribution of Impacts of Natural Disasters Across Income Groups: A Case Study of New Orleans," *Ecological Economics* 63 (2007): 299–306.

188n *Climate change has already made*: Noah S. Diffenbaugh and Marshall Burke, "Global Warming Has Increased Global Economic Inequality," *PNAS* 116 (2019): 9808–13.

189 *For the last two centuries*: Information about Isle de Jean Charles, and English's work with the community there, is from: "Isle de Jean Charles," Isle de Jean Charles, Louisiana, accessed May 24, 2019, www.isledejeancharles .com/. "Ghost Trees and Legends," Isle de Jean Charles, Louisiana (blog), June 7, 2017, www.isledejeancharles.com/new-blog/2017/6/6/ghost-trees -and-legends. "Tribal Resettlement," Isle de Jean Charles, Louisiana, accessed May 24, 2019, www.isledejeancharles.com/our-resettlement/. The Isle de Jean Charles Biloxi-Chitimacha-Choctaw Tribe and Tribal Council, "The Isle de Jean Charles Tribal Resettlement: A Tribal-Driven, Whole Community Process," news release, January 15, 2019, www.isledejeancharles.com/s/IDJC-Press-release-1-18-19.pdf. English, Klink, and Turner, "Thriving with Water." English, interview, August 8, 2018.

189 *In addition, she and her colleagues . . . safe from rising floodwaters*: English, Klink, and Turner, "Thriving with Water." English, interview, August 2, 2017.

"Projects," Buoyant Foundation Project, accessed May 24, 2019, http://buoyantfoundation.org/work/projects/.

190 *In 2018, the Buoyant Foundation Project . . . scale up the project*: English, interview, August 15, 2017. English, interview, August 8, 2018. Elizabeth English, e-mail to author, May 1, 2019. "Amphibious Retrofitting in the Mekong River Delta, Vietnam," Buoyant Foundation Project, accessed May 24, 2019, http://buoyantfoundation.org/amphibious-retrofits-in-the-mekong-river-delta/. "Flood Resilient Homes in Vietnam and Bangladesh," Global Resilience Project, August 27, 2018, video, www.youtube.com/watch?v=ATYoUF9XI-A.

191 *The Dutch architecture firm*: "Projects," Waterstudio, accessed May 30, 2019, www.waterstudio.nl/projects/.

191 *In 2019, the company Oceanix*: Ben Guarino, "As Seas Rise, the U.N. Explores a Bold Plan: Floating Cities," *The Washington Post*, April 5, 2019, www.washingtonpost.com/science/2019/04/05/seas-rise-un-explores-bold-plan-floating-cities/. "Oceanix," Oceanix, accessed May 24, 2019, https://oceanix.org/.

191 *In the United States, federal law*: Fenuta, "Amphibious Architectures," 138–52. English, interview, August 2, 2017. English, interview, August 15, 2017.

191 *In 2007, an official*: Fenuta, "Amphibious Architectures," 138–39. English, interview, August 2, 2017.

191 *"Although amphibious building technology"*: FEMA Press Office, e-mail to author, August 17, 2017.

191 *She received a grant*: English, interview, August 2, 2017. English, interview, August 8, 2018.

192 *In Joplin, Missouri*: "CORE: Disaster Resilient Design," Q4 Architects, accessed May 24, 2109, http://q4architects.com/projects/core-disaster-resilient-design/. Emily Badger, "An Ingenious Home Built to Battle Tornadoes," *CityLab*, October 3, 2013, www.citylab.com/equity/2013/10/ingenius-home-built-battle-tornadoes/7105/.

192 *Deltec Homes, a North Carolina–based*: "Features and Options," Deltec, accessed May 24, 2019, www.deltechomes.com/features-options/. "This Hurricane-Proof Home Can Withstand Powerful Storms, Thanks to Its Aerodynamic Design," *The Verge*, November 8, 2017, www.theverge.com/2017/11/8/16619006/deltec-hurricane-proof-house-harvey-irma-maria-home-of-the-future.

193 *After a major earthquake*: "40 Buildings Fell During Earthquake Due to Corruption, Organization Charges," *Mexico News Daily*, September 11, 2018, https://mexiconewsdaily.com/news/buildings-fell-during-earthquake-due

-to-corruption/. Martha Pskowski, "Mexico City's Architects of Destruction," *CityLab*, September 19, 2018, www.citylab.com/environment/2018/09/mexico-city-earthquake-damage-building-codes/570679/.

193 *Building just one new*: *2011 Buildings Energy Data Book* (U.S. Department of Energy, 2012), 1–27. Alex Wilson and Jessica Boehland, "Small Is Beautiful: U.S. House Size, Resource Use, and the Environment," *Journal of Industrial Ecology* 9 (2005): 277–87.

193 *The construction and operation*: *Towards a Zero-Emission, Efficient, and Resilient Buildings and Construction Sector: Global Status Report 2017* (United Nations Environment and International Energy Agency, 2017), 14.

194 *Seattle's Bullitt Center*: Information about the Bullitt Center is from: "Building Features," Bullitt Center, accessed May 24, 2019, www.bullittcenter.org/building/building-features/. "Living Building Challenge," Bullitt Center, accessed May 24, 2019, www.bullittcenter.org/vision/living-building-challenge/. Bullitt Center, "Bullitt Center Earns Living Building Certification," news release, April 1, 2015, www.bullittcenter.org/2015/04/01/bullitt-center-earns-living-building-certification/. Bullitt Center, "Bullitt Center: A Project of the Bullitt Foundation," media kit, November 2013, www.bullittcenter.org/field/media/media-kit/.

194 *Powerhouse, a consortium*: "Projects," Powerhouse, accessed May 24, 2019, www.powerhouse.no/en/projects/. Tracey Lindeman, "Norway's Energy-Positive Building Spree Is Here," *CityLab*, December 13, 2018, www.citylab.com/environment/2018/12/norway-energy-positive-building-powerhouse-snohetta/577918/.

194 *Sidewalk Labs . . . waste collection robots*: Charlotte Matthews, "Creating a Pathway to Climate Positive Communities, *Medium*, https://medium.com/sidewalk-toronto/creating-a-pathway-to-climate-positive-communities-32b67c85d528. *Quayside Site Plan*, draft (Sidewalk Toronto, November 29, 2018), https://sidewalktoronto.ca/wp-content/uploads/2018/11/18.11.29_Quayside_Draft_Site-Plan.pdf. *Sustainability* (Sidewalk Labs, 2019), https://sidewalktoronto.ca/wp-content/uploads/2019/01/DRP_Sustainability.pdf.

194 *Societies that settled in . . . deflect high-speed winds*: Sandra Piesik, ed., *Habitat: Vernacular Architecture for a Changing Planet* (New York: Harry N. Abrams, 2017).

195 *some in seismically active zones*: Wen-bao Luo, "Seismic Problems of Cave Dwellings on China's Loess Plateau," *Tunnelling and Underground Space Technology* 2 (1987): 203–208. Gideon Golany, *Chinese Earth-Sheltered Dwellings: Indigenous Lessons for Modern Urban Design* (Honolulu: University of Hawaii Press, 1992), 138.

195 *Khalili was born in 1936*: Information about Khalili's life and work is from my own visit to CalEarth on March 3, 2018, as well as: Sheefteh Khalili, interview by author, January 8, 2018. Sheefteh Kahlili, e-mail to author, April 10, 2019. Nader Khalili, *Racing Alone*, 4th ed. (Hesperia, CA: CalEarth Press, 2003). *Sandbag Shelter Prototypes* (Aga Khan Award for Architecture, 2004), www.akdn.org/architecture/project/sandbag-shelters. "Our Founder," CalEarth, accessed May 25, 2019, www.calearth.org/our-founder. "SuperAdobe: Powerful Simplicity," CalEarth, accessed May 25, 2109, www.calearth.org/intro-superadobe.

196 *"My little boy . . . to race alone"*: Khalili, *Racing Alone*, 48.

196 *"The right to life cannot be separated"*: *Adequate Housing as a Component of the Right to an Adequate Standard of Living*, A/71/310 (New York: United Nations, 2016), 11/24, https://digitallibrary.un.org/record/840297.

196 *And yet, more than a billion*: *Report of the Special Rapporteur on Adequate Housing as a Component of the Right to an Adequate Standard of Living, Miloon Kothari*, E/CN.4/2005/48 (United Nations Commission on Human Rights, 2005), 2.

196 *with hundreds of millions of people*: Madhu Thangavelu, "Lunar and Terrestrial Sustainable Building Technology in the New Millennium: An Interview with Architect Nader Khalili," *Building Standards* (2000): 44–47.

197 *developed a system*: Information about SuperAdobe is from my own visit to CalEarth on March 3, 2018, as well as: Khalili, interview, January 8, 2018. *Sandbag Shelter Prototypes*. "SuperAdobe: Powerful Simplicity," CalEarth. "At CalEarth," CalEarth, accessed May 25, 2019, www.calearth.org/superadobe-structures-calearth.

197 *"The foundations are buried"*: Khalili, *Racing Alone*, 28.

198 *"In fact, if we hadn't . . . were greatly exceeded"*: Tom Harp and John Regner, "Sandbag/Superadobe/Superblock: A Code Official Perspective," *Building Standards* (1998): 28.

198 *which he patented in 1999*: Ebrahim Nader Khalili, "Earthquake Resistant Building Structure Employing Sandbags," U.S. Patent 5,934,027, filed February 19, 1998, and issued August 10, 1999.

198 *SuperAdobe's first real-world . . . in less than two weeks*: *Sandbag Shelter Prototypes*, 20–24.

199 *"Sandbags and barbed wire"*: *Sandbag Shelter Prototypes*, 22.

199n *Five years after the quake*: International Organization for Migration, "Five Years After 2010 Earthquake, Thousands of Haitians Remain Displaced," news release, January 9, 2015, www.iom.int/news/five-years-after-2010-earthquake-thousands-haitians-remain-displaced.

201 *There are now SuperAdobe structures*: "SuperAdobe Worldwide," CalEarth, accessed May 26, 2019, www.calearth.org/alumni-projects2.

201 *In 2015, a forty-dome*: Khalili, interview, January 8, 2018. CalEarth, "Superadobe/Earthbag Orphanage Withstands Nepal Earthquake," news release, May 5, 2015, www.calearth.org/blog/2016/6/23/for-immediate -release-may-5-2015-superadobeearthbag-orphanage-withstands-nepal -earthquake.

201 *Two SuperAdobe domes*: Sheefteh Khalili, e-mail to author, May 10, 2019.

201 *In December 2017 . . . turned to ash*: Khalili, interview, January 8, 2018. "Fire Update #3: December 31, 2017," The Ojai Foundation, December 31, 2017, https://ojaifoundation.org/news-and-updates/thomas_fire_update_03/.

201 *And the institute . . . permits for SuperAdobe buildings*: Khalili, interview, January 8, 2018. Khalili, e-mail, April 10, 2019. "Help Save CalEarth: De-tails and Timeline," CalEarth, accessed May 26, 2019, www.calearth.org /timeline. "CalEarth has BIG News to Share!" CalEarth (blog), February 5, 2019, www.calearth.org/blog/2019/2/5/calearth-has-big-news-to-share.

202 *In January 2018*: Khalili, interview, January 8, 2018.

202 *Japan Dome House*: "Superior Features of Dome House," Japan Dome House, accessed May 26, 2019, www.i-domehouse.com/page02.html. "Proven and Certified Reliable Performance," Japan Dome House, accessed May 26, 2019, www.i-domehouse.com/page03.html. "Japan's Earthquake-Resistant Dome Houses," Reuters, November 6, 2017, www.reuters.com/news/picture /japans-earthquake-resistant-dome-houses-idUSRTS1IP2W.

202 *Meanwhile, some American organizations*: Alex Wilson, "Resilient Tiny House Shelters for the Homeless," April 4, 2018, www.resilientdesign.org /resilient-tiny-house-shelters-for-the-homeless/.

202 *In 2018, ICON . . . for impoverished families*: ICON, "New Story and ICON Unveil the First Permitted 3D-Printed Home," news release, March 15, 2018, www.iconbuild.com/updates/this-house-can-be-3d-printed-for-cheap. "Fre-quently Asked Questions," ICON, accessed May 26, 2019, www.iconbuild .com/about/faq.

202 *To fashion more sustainable structures*: Ad van Wijk and Iris van Wijk, *3D Printing with Biomaterials: Towards a Sustainable and Circular Economy* (Amsterdam: IOS Press, 2015). "Biomaterials for Additive Manufacturing," Oak Ridge National Laboratory (blog), March 23, 2018, www.ornl.gov/blog /eesd-review/biomaterials-additive-manufacturing. Patricia Leigh Brown, "The Lewis and Clark of the Digital Building Frontier," *The New York Times*, March 15, 2019, www.nytimes.com/2019/03/15/arts/design/3d -printing-building-design.html.

9. BLUEPRINTS FOR THE RED PLANET

205 *"There are even fewer"*: Thangavelu, "Lunar and Terrestrial."

205 *In 1984, Khalili presented*: E. Nader Khalili, "Magma, Ceramic, and Fused Adobe Structures Generated *In Situ*," in *Lunar Bases and Space Activities of the 21st Century*, ed. W. W. Mendell (Lunar and Planetary Institute, 1985), 399–403.

206 *"a centrifugally gyrating platform"*: Khalili, "Magma, Ceramic, and Fused."

206 *In the years after*: Thangavelu, "Lunar and Terrestrial."

206 *It was in those years*: Khalili, interview, January 8, 2018. E. Nader Khalili, "Lunar Structures Generated and Shielded with On-Site Materials," *Journal of Aerospace Engineering* 2 (1989): 119–29.

206 *"Discovering suitable dimensions"*: Khalili, "Magma, Ceramic, and Fused," 402.

207 *Sherwood was born*: Information about Sherwood's background and work is from: Brent Sherwood, interview by author, September 15, 2017.

207 *"a sustained human presence"*: "Blue Moon," Blue Origin, www.blueorigin.com/blue-moon, accessed July 17, 2019.

209n *Some scientists dream*: A. J. Berliner and C. P. McKay, "The Terraforming Timeline" (paper, Planetary Science Vision 2050 Workshop, Washington, DC, February 27–March 1, 2017).

210 *Or we could swap out*: Houssam A. Toutanji, Steve Evans, and Richard N. Grugel, "Performance of Lunar Sulfur Concrete in Lunar Environments," *Construction and Building Materials* 29 (2012): 444–48. Violeta Gracia and Ignasi Casanova, "Sulfur Concrete: A Viable Alternative for Lunar Construction," *Space* 98 (1998): 585–91.

210 *Engineers at Stanford*: Edmund L. Andrews, "A New Technique Could Help Turn Mars or Moon Rocks into Concrete," *Stanford Engineering Magazine*, May 2, 2017, https://engineering.stanford.edu/magazine/article/new-technique-could-help-turn-mars-or-moon-rocks-concrete.

210 *In 2013, the European Space Agency*: "Building a Lunar Base with 3D Printing," European Space Agency, January 31, 2013, www.esa.int/Our_Activities/Space_Engineering_Technology/Building_a_lunar_base_with_3D_printing.

211 *In 2017, ESA researchers*: "Printing Bricks from Moondust Using the Sun's Heat," European Space Agency, May 3, 2017, www.esa.int/Our_Activities/Space_Engineering_Technology/Printing_bricks_from_moondust_using_the_Sun_s_heat.

211 *In 2015, NASA*: "3D-Printed Habitat Challenge," NASA, accessed May 26, 2019, www.nasa.gov/directorates/spacetech/centennial_challenges

/3DPHab_p1.html. "NASA's Centennial Challenges: 3D-Printed Habitat Challenge," NASA, accessed May 26, 2019, www.nasa.gov/directorates/spacetech/centennial_challenges/3DPHab/about.html.

211 *The team that won*: "Mars Ice House," Mars Ice House, accessed May 26, 2019, www.marsicehouse.com/.

211 *Another team proposed*: "Marsha," AI SpaceFactory, accessed May 26, 2019, www.aispacefactory.com/marsha. "AI SpaceFactory - MARSHA - Our Vertical Martian Future - Part One," AI SpaceFactory, July 23, 2018, video, www.youtube.com/watch?v=XnrVV0w2jrE.

211 *one of its most ambitious concepts*: Information about TransHab is from: "TransHab Concept," NASA, accessed May 26, 2019, https://spaceflight.nasa.gov/history/station/transhab/. "TransHab Concept," NASA, accessed May 26, 2019, https://spaceflight.nasa.gov/history/station/transhab/transhab_levels.html.

214 *In 2016, NASA replaced*: "The Power of Light," NASA, accessed May 26, 2019, https://science.nasa.gov/news-articles/the-power-of-light. "Testing Solid State Lighting Countermeasures to Improve Circadian Adaptation, Sleep, and Performance During High Fidelity Analog and Flight Studies for the International Space Station," NASA, accessed May 26, 2019, www.nasa.gov/mission_pages/station/research/experiments/explorer/Investigation.html?#id=2013.

214 *The Soviet space program . . . more time gardening*: Sandra Häuplik-Meusburger et al., "Greenhouses and Their Humanizing Synergies," *Acta Astronautica* 96 (2014): 138–50.

215 *One of the teams*: "Alpha," Mars City Design, accessed May 26, 2019, www.marscitydesign.com/alpha.

216 *In a 2011 paper*: Ayako Ono and Irene Lia Schlacht, "Space Art: Aesthetics Design as Psychological Support," *Personal and Ubiquitous Computing* 15 (2011): 511–18.

216 *"All the conditions necessary"*: The quote appears in Angel Marie Seguin, "Engaging Space: Extraterrestrial Architecture and the Human Psyche," *Acta Astronautica* 56 (2005): 980–95.

216 *In August 2015, a crew of six*: Information about the mission is from: Sandra Häuplik-Meusburger, Kim Binsted, and Tristan Bassingthwaighte, "Habitability Studies and Full Scale Simulation Research: Preliminary Themes Following HISEAS Mission IV," paper presented at the 47th International Conference on Environmental Systems, Charleston, SC, July 2017. Sandra Häuplik-Meusburger, interview by author, April 10, 2019.

217 *"Activities which involved"*: Häuplik-Meusburger, Binsted, and Bassingthwaighte, "Habitability Studies."

219 *NASA has already developed*: "High-Efficiency Solar Cell," NASA, accessed May 26, 2019, https://technology.nasa.gov/patent/LEW-TOPS-50. "Living Blue," NASA, accessed May 26, 2019, www.nasa.gov/ames/facilities /sustainabilitybase/livingblue.

219 *In 2017, that possibility . . . small apartments*: "The IKEA Journey into Space Just Started," IKEA, September 6, 2017, http://ikea.today/ikea-journey-space-just-started/. IKEA, "RUMTID," news release, June 7, 2018, https:// newsroom.inter.ikea.com/events/rumtid/s/a22b73d8-fd5d-47c4-981e -03d06e3db9cc. Jeremy White, "IKEA Designers Are Living in a Mars Simulator to Get Inspiration for Future Collections. Really," *Wired UK*, June 8, 2017, www.wired.co.uk/article/ikea-and-mars. Katharine Schwab, "What Ikea's Designers Learned from Living in a Simulated Mars Habitat," *Fast Company*, June 20, 2017, www.fastcompany.com/90130253/what-ikeas -designers-learned-from-living-inside-a-mars-simulation. Aileen Kwun, "See the Collection Ikea Designed for Tiny Apartments—by Studying Mars," *Fast Company*, June 14, 2018, www.fastcompany.com/90175873/ikeas-latest -collection-involved-living-in-a-mars-simulator.

ACKNOWLEDGMENTS

This was a difficult book to write, and it took me far longer than I anticipated. I was fortunate to have lots of help along the way. I couldn't have done any of it without Abigail Koons, the world's most patient agent. She has talked me off many ledges and is always in my corner. My editor, Amanda Moon, was enthusiastic about this idea from day one and helped me shape my sprawling, shaggy first draft into a coherent story. It was a joy to work with her again. Colin Dickerman provided a fresh eye as the manuscript neared completion and helped usher it across the finish line. I appreciate his hard work and that of everyone else at FSG, which has been a wonderful publishing partner.

A number of friends and family members provided feedback on

early versions of the manuscript. I'm grateful to Gary Anthes, Blaine Boman, Jessica Feinstein, Brian Ha, Melanie Loftus, Caroline Mayer, Ben Plotz, Michelle Sipics, Nick Summers, and Jieun Yang for helping to save me from my own mistakes.

I am indebted to all the scientists, architects, researchers, and other sources who generously shared their stories and experiences with me. Many of them do not appear in these pages—I interviewed far more people than I had room to include—but they all provided crucial information and perspective that helped shape my research and thinking. Across the board, they were generous with their time and gracious in the face of what must have sometimes seemed like an endless stream of follow-up questions.

Finally, for their unflagging faith and support, in writing and in life, thanks and love to Gary, Caroline, Ali, and Blaine.

INDEX

cardiovascular disease, 60–61, 69
cars, 59–60
Carter, Stephen, 145, 150, 152, 154*n*
Carter Goble Lee (CGL), 144–45, 150
Casa Anfibia, 190
CDC, *see* Centers for Disease Control and Prevention
Center for Active Design, 62–64
Center for Care and Discovery, 31
Center for Health Design, 42
Center for Health Facilities Design and Testing, 45
Center for the Built Environment, 100*n*
Centers for Disease Control and Prevention (CDC), 63*n*; physical activity recommendations, 64
Centre for Urban Design and Mental Health, 155
CGL, *see* Carter Goble Lee
chemical disinfection, 36
children, active design and, 64–65
chocolate milk, 71–72
cholera, 28, 55–57
chronic disease, 10, 57, 79; in senior care residents, 164
circadian lighting, 40, 89
circadian rhythms, 40
circulating nurses, 46, 48–50
Citizens' Association of New York, 56
civil rights, 105
clean water, access to, 28
Clemson Design Center, 45, 49
Clemson University, 45, 52
climate chambers, 84

climate change, 9, 182, 188, 192–94
Clorox, 174
Clostridium difficile, 32
cockroaches, 22
cognitive disabilities, 106, 108; design for, 117–18
cognitive performance: nature and, 89; visual complexity of rooms and, 8
cognitive variation, 108
Colistra, Joe, 172–74, 176
Comfy, 100
common spaces, 112–13
communication: digital, 92–93; face-to-face, 91–93; public spaces and, 156; workplace patterns of, 91
community commons areas, 73, 119, 120
conference rooms, 95
control units, 135
coronary heart disease, 69
correctional facilities, 131, 133; humane, 143; normalization and, 146; vocational training in, 145, 149
correctional system, 136–37
Corynebacterium, 17
Council of Residential Engagement, 126
co-working, 93
CPAP machines, 174
Crane, Sam, 106, 107, 124, 125
Crimean War, 35
criminal justice system, reform efforts for, 142–43, 153
cross-ventilation, 36

McDonnell Douglass Space Systems, 206

Medical University of South Carolina (MUSC), 45, 46, 52

meeting rooms, 95, 96

Mekong Delta, 190

mental health: of astronauts, 215; building design and, 64*n*; built environment and, 131; criminal justice reform and, 153; isolation and, 142; prison facilities for, 149, 153; solitary confinement and, 138; urban design and, 155, 156; windows and, 6

mental hospitals, 154

mental illness, 106; prison populations and, 143; prisons and, 131; solitary confinement and, 138

messaging software, 91

Metropolitan Board of Health, 56

micro-apartments, 194

microbes, 12; beneficial, 29; benefits of exposures to, 24–25; dampness and growth of, 20; in hospitals, 31–32; indoor, 17, 24; skin, 17

microbiology, 14, 15

microbiomes, 16, 30; indoor, 19, 28

microorganisms, 12

Microsoft Kinect, 163, 168

Mitaka lofts, 3–5

MLE1, 26

mobility, architecture and planning removing, 59–60

mold, 24

monsoons, 183

mood: imprisonment and, 138; outdoor landscapes effects on, 37

Moon: construction on, 206, 207; day and night cycles on, 214; environmental challenges of, 208–209; settlements on, 208

Morris, Jerry, 69

mortality, 4

mosquitoes, 23

moss, 23

Mostafa, Magda, 108

motion-capture cameras, 163

motion detector systems, 171

Mouchka, Randy, 82

Mount Sinai, 63, 64

movement, designed out of cities, 59–60

movement temptations, 68

Mulyani, Vera, 213, 215

MUSC, *see* Medical University of South Carolina

Musk, Elon, 208

mycobacteria, 27

N

NASA, 205–208, 211, 214, 215, 219

National Flood Insurance Program (NFIP), 191

National Institute of Child Health and Human Development, 65

National Research Council, 191

National Restaurant Association, 149

natural disasters, 183, 193

natural scenery, 37, 43; benefits of exposure to, 39–40; in prisons, 147; views of, 38, 39

resilience, forms of, 192–93

resilient buildings, 192

Resnik, Denise, 110, 113, 114, 118–23, 125, 127

Resnik, Matt, 110, 111, 127

reversible destiny, 4

Reversible Destiny Foundation, 4, 5

Rickettsia, 22

Rikers Island, 130, 152

Robert Wood Johnson Foundation, 64

Rocky Mountain spotted fever, 22

RoCo, 100*n*

Roseburia, 17

RSP Architects, 113

S

Safe OR Design Tool, 52

Salyut 1 space station, 215

San Diego County sheriff's department, 144

sanitary reform, 56

SARRC, *see* Southwest Autism Research and Resource Center

SARS, 32

scattered-site housing, 124

Schnädelbach, Holger, 178

schools, 65–80; cafeterias, 70–71, 74, 76; exercise and, 66, 68, 69; learning streets in, 68; nutrition education in, 71, 73–74, 77; old buildings, 65–66; outdoor areas, 72–73; physical activity and, 77–78; renovating, 67–68; standing and dynamic furniture in, 70; student self-perception

and, 155; walking in, 75; wellness programs in, 74

Scialla, Paul, 83

scrub nurses, 50, 52

seizures, 111

senior care, 161, 169–70; dehumanizing products for, 177; fall detection for, 163–64

senior homes, 118

Senseable City Lab, 100

sensory deprivation, 139

sensory experience: in prisons, 141; of solitary confinement, 140, 141; visual complexity and, 141

sensory inclusion, 110

sensory monotony, 140

sensory overload, minimizing in living spaces, 112

sensory rooms, 112

sensory stimuli, autism and, 110, 112, 115–17

sepsis, 10

sewers, 55–56

Sherwood, Brent, 206–207, 210, 213, 217, 218

Shimberg Center for Housing Studies, 110

showerheads, 11–13, 26, 27

showers, bacteria and, 13

Sidewalk Labs, 194

Silverstein, Thomas, 135

Simon, Madlen, 122

Sinclair Home Care, 162

sitting, 69–70

Skåne University Hospital, 32–34

skin microbes, 17

A NOTE ABOUT THE AUTHOR

Emily Anthes is an award-winning science journalist and author. Her work has appeared in *The New York Times*, *The New Yorker*, *The Atlantic*, *Wired*, *Nature*, *Slate*, *Bloomberg Businessweek*, *Scientific American*, *The Washington Post*, *The Boston Globe*, and other publications. Her previous book, *Frankenstein's Cat*, was long-listed for the PEN/E. O. Wilson Literary Science Writing Award.